Advances in Principal Component Analysis

Edited by Fausto Pedro García Márquez

Published in London, United Kingdom

Advances in Principal Component Analysis
http://dx.doi.org/10.5772/intechopen.97992
Edited by Fausto Pedro García Márquez

Contributors
Laura-Maria Dogariu, Constantin Paleologu, Silviu Ciochină, Jacob Benesty, Florian De Vuyst, Anne-Virginie Salsac, Claire Dupont, Souhaila Chahboun, Mohamed Maaroufi, Stanley L. Sclove, Zaloa Sanchez-Varela, Branko Perišić, Ana Perišić, Stefana Janicijevic, Maja Kljajić, Vule Mizdraković, Kam Tim Tse, Lei Zhou, Hiroko Katayama, Yuichi Mori, Masahiro Kuroda, Padmavathi Ganapathi, Roshni Arumugam, Shanmugapriya Dhathathri, Shouichi Takane, Augustine Iduseri

First published in London, United Kingdom, 2022 by IntechOpen
IntechOpen is the global imprint of INTECHOPEN LIMITED, registered in England and Wales, registration number: 11086078, 5 Princes Gate Court, London, SW7 2QJ, United Kingdom

British Library Cataloguing-in-Publication Data
A catalogue record for this book is available from the British Library

Additional hard and PDF copies can be obtained from orders@intechopen.com

Advances in Principal Component Analysis
Edited by Fausto Pedro García Márquez
p. cm.
Print ISBN 978-1-80355-765-6
Online ISBN 978-1-80355-766-3
eBook (PDF) ISBN 978-1-80355-767-0

For EU product safety concerns:
IN TECH d.o.o., Prolaz Marije Krucifikse Kozulić 3, 51000 Rijeka, Croatia,
info@intechopen.com or visit our website at intechopen.com.

Meet the editor

Fausto Pedro García Márquez is a full professor at the University of Castilla–La Mancha (UCLM), Spain, an honorary senior research fellow at Birmingham University, UK, and a lecturer at the Postgraduate European Institute. In 2013/2014, he was a senior manager at Accenture. He obtained his European Ph.D. with maximum distinction. He is the recipient of several awards from various organizations, including the International Society of Management Science and Engineering Management (ICMSEM). He has published more than 170 journal papers and authored and edited 40 books, 90 book chapters, and 6 patents. He is an editor for five international journals and a committee member of more than sixty conferences. He has been the Principal Investigator of more than 150 projects both national and international. His main interests include artificial intelligence, maintenance, management, renewable energy, transport, advanced analytics, and data science.

Contents

Preface **IX**

Chapter 1 **1**
The Foundation for Open Component Analysis: A System
of Systems Hyper Framework Model
by Ana Perišić and Branko Perišić

Chapter 2 **27**
Identification of Multilinear Systems: A Brief Overview
*by Laura-Maria Dogariu, Constantin Paleologu, Jacob Benesty
and Silviu Ciochină*

Chapter 3 **47**
Evaluation of Principal Component Analysis Variants
to Assess Their Suitability for Mobile Malware Detection
*by Padmavathi Ganapathi, Shanmugapriya Dhathathri
and Roshni Arumugam*

Chapter 4 **67**
Principal Component Analysis and Artificial Intelligence
Approaches for Solar Photovoltaic Power Forecasting
by Souhaila Chahboun and Mohamed Maaroufi

Chapter 5 **79**
Variable Selection in Nonlinear Principal Component Analysis
by Hiroko Katayama, Yuichi Mori and Masahiro Kuroda

Chapter 6 **91**
Space-Time-Parameter PCA for Data-Driven Modeling
with Application to Bioengineering
by Florian De Vuyst, Claire Dupont and Anne-Virginie Salsac

Chapter 7 **113**
Principal Component Analysis in Financial Data Science
by Stefana Janićijević, Vule Mizdraković and Maja Kljajić

Chapter 8 139
Determining an Adequate Number of Principal Components
by Stanley L. Sclove

Chapter 9 155
Spatial Principal Component Analysis of Head-Related Transfer Functions and Its Domain Dependency
by Shouichi Takane

Chapter 10 175
Prediction Analysis Based on Logistic Regression Modelling
by Zaloa Sanchez-Varela

Chapter 11 199
On the Use of Modified Winsorization with Graphical Diagnostic for Obtaining a Statistically Optimal Classification Accuracy in Predictive Discriminant Analysis
by Augustine Iduseri

Chapter 12 219
Mode Interpretation of Aerodynamic Characteristics of Tall Buildings Subject to Twisted Winds
by Lei Zhou and Kam Tim Tse

Preface

Nowadays, the use of large amounts of data in industry is increasing because of the development of new sensors and more complex systems, which require increased reliability, availability, and safety. The Internet of Things (IoT) is one example of a source that creates large amounts and varieties of data.

Principal component analysis (PCA) is a statistical procedure increasingly being applied to analyze large datasets. It is mainly used to modify datasets, reducing the coordinate system by a linear transformation. The new, smaller set of information extracted via PCA is a set of summary indices called "principal components." The way to obtain these principal components is according to variance, where the first principal component is obtained for the greater variance.

The main objective of PCA is to transform the size of a dataset into a smaller transformed space provided by the eigenvectors of the covariances associated to the original dataset. A ranking of each eigenvector is done following the maximum variability. These are called principal components. In other words, the method transforms the original dataset into a new p-dimensional set of Cartesian coordinates, being a projection of the dataset onto the principal component vector, where the direction is given by the P matrix, where a is the first largest eigenvalue and its columns are the retained eigenvectors. PCA can be also related by canonical correlation analysis, where the coordinate systems are obtained optimally by the cross-covariance considering two datasets together, while PCA defines a new orthogonal coordinate system that optimally describes the variance in a single dataset.

PCA can also be associated with other algorithms, but with some differences, for example, factor analysis, non-negative matrix factorization, correspondence analysis, K-means clustering, and so on. PCA does have some shortcomings and thus generalizations of the technique have been developed to overcome these limitations. These include sparse PCA, robust PCA, and nonlinear PCA.

This book provides a comprehensive overview of PCA and its analytical principles. It examines the use of PCA in a variety of fields, including technology, engineering, finance, risk analysis, marketing, economics, and more. It presents practical case studies highlighting the use of PCA in several types of industries to solve problems both small and large using simple and complex algorithms.

Fausto Pedro García Márquez
Ingenium Research Group,
University of Castilla-La Mancha,
Ciudad Real, Spain

Chapter 1

The Foundation for Open Component Analysis: A System of Systems Hyper Framework Model

Ana Perišić and Branko Perišić

Abstract

The interoperability and integration of heterogeneous systems, with a high degree of autonomy and time-dependent dynamic configuration over multilevel and multidimensional feature space, raise the problem configurations complexity. Due to the emergent nature of a large collection of locally interacting components, the properties and the behavior of a collection may not be fully understood or predicted even the full knowledge of its constituents is available. The simplification is contemporary addressed through either dimensional reduction methods, like Principal Component Analysis (PCA), or overall ontology managing through Physics of Open Systems (POS) paradigm. The question is: Is it possible to cope with the complexity by integrating dimension reduction steps with basic POS concepts on the Large Data Objects (LDOs) holding the structure and behavior of the complex system. The intended mission of this chapter is to formulate a starting System of Systems (SoS) based configurable hyper framework model that may be dynamically improved to better suit the static structure and dynamic behavior of complex SoS configurations. That is the reason why the reflexive integration of POS and different dimensional reduction methods, through an interoperability framework, have been proposed as the main contribution of this research chapter.

Keywords: collections complexity, framework modeling, large data objects, principal component analysis, physics of open systems, heterogeneous systems interoperability, system of systems analysis

1. Introduction

The globally accepted definitions of digitalization and digital transformation do not still exist, although the terms are in the field for quite a long time. In Gartner Glossary, digitalization is defined as *the use of digital technologies to change a business model and provide new revenue and value-producing opportunities; it is the process of moving to a digital business* [1]. This definition accents the higher granularity mission concerning global system aspects. Considering the particular enterprise systems, that have decided to move their business into the digital form, there is a challenging activity of developing a completely new set of processes and procedures in compliance with the formulated digital business model. This process is usually considered as enterprise digital transformation. The proliferation of Information and Communication Technologies (ICT), especially in the last decade, puts digitalization and digital transformation into the focus of arbitrary systems development and sustainable collaboration/cooperation in the context of a huge number of dynamic configurations that may emerge throughout its entire life cycle.

The interoperability of heterogeneous systems and integration processes favor System of Systems (SoS) analysis and synthesis approaches to dominate through the 21st century, a century that has become the era of complexity. From the architectural point of view, every real-life system is an SoS, where its architecture is usually seen as a mechanism that creates the illusion of simplicity. To avoid the traditional hierarchical approach to the representation of systems internals, the SoS metaphor has to be deeply understood due to its inherent meshed topology. SoS may range from the technical-systems counterparts, dominantly addressed in systems engineering, to the social systems like, for example, the education, teaching, learning, and performing ecosystem that favor personal competency profile, that suits the industry 4.0 competencies compliance. An integrated systems approach needs to provide a framework and language that allow different systems, with highly divergent characteristics, to interoperate in favor of the commonly agreed mission. With the natural multidimensionality, embedded in the generic SoS paradigm, the underlining complexity directly affects the management and control of the resulting SoS. As a consequence, the dynamic multilayered architecture emerges as a promising approach to the internals complexity hiding.

The SoS concept assumes that participating systems are characterized by the: autonomous mission; independence of encapsulated operations; the difference in fundamental life cycle model aspects. The SoS life cycle stages generally include *creation* (that usually strictly separates the creation process from operational usage); *sustainable operation* (leaving the variety of SoS artifacts persistent instances); *migration* (SoS evolution that retains functionality and structure while replacing the supportive technologies); *replacement* (complete or partial component replacement while preserving the interoperability over specified interfaces); and *termination* (the retirement of a component(s) system(s) while preserving the structural and functional consistency of the remaining SoS). The SoS level of integration is dominantly differentiated by the way the participating systems are orchestrated. According to ref. [2] SoS may be divided into four main categories: *centrally* (directed); *cooperatively* (acknowledged); *collaboratively* (with common enforcing mechanisms); *emergency control* (virtual control of uncontrolled large-scale behavior) *orchestrated*. To cope with the complexity, the architectural design tends to hide the internals of complex systems by wrapping them with an adaptation layer that, to the outer layer stakeholders, exhibits only high-level granularity concepts through well-defined interfaces.

In general, complex systems are analyzed from two complementary aspects: the structure (static view) and the behavior (dynamic view). The structural complexity emerges from observing the system as a composition of arbitrary, interrelated components (subsystems). These components may be considered or sometimes really are systems with a high degree of autonomy. From the behavioral aspect, complexity is defined as the degree of difficulty in predicting the future dynamic properties and behavior of an overall system assuming that current dynamic properties and behavior of the participating components (subsystems) are known. Due to the emergent nature of a large collection of locally interacting components, the properties and the behavior of a collection may not be fully understood or predicted even the full knowledge of its constituents is available. Because of that, Complex Systems Science (CSS) requires new mathematical and heuristic frameworks and scientific methodologies to cope with SoS organized structural and behavioral complexity handling.

The complementary approach is to view an arbitrary system through its inherent dimensions, which do not strictly structure or behavior-oriented but add another direction of complexity, the multidimensionality. Combined with the multilevel architecture it creates multidimensional multilevel (hyper) composites. If there is a need for different contexts or configurations management in timed framed scale, the hyper composites may generate an arbitrarily large number of instances that are context, configuration, and time-dependent. These instances form a generic repository that is usually seen as an association of Large Data Objects (LDOs). These objects need to be stored, loaded, searched, modified, and visualized in vide variety of ways. Traversing, for example, the hyper-structure instances, with any rational reason in mind, creates a challenging NP (nondeterministic polynomial time) complex problem that opens the fundamental question of arbitrary dimensionality reduction approach. The dimensionality reduction facilitates the transformation of the initial system/problem into a less complex one (with a lower degree of freedom) that may be efficiently/effectively solved with acceptably lower accuracy (precision).

Modeling is considered a promising approach to cope with complexity. Model-driven SoS development is a challenging paradigm that may generate the initial multidimensional and multilayered meta-SoS model (MSoSM). The originating version of any particular MSoSM may serve as a starting orchestration pattern that, by dynamical inclusion of individual dimensions in particular configuration(s), forms meta structures and facilitates their instantiation. According to that, one of the main challenging approaches, being the main goal of this chapter, is the specification of SoS based open framework model that may serve as the meta-SoS pattern for describing and monitoring arbitrary complex composites that instantiate multilayered multidimensional data objects that may serve as a resources pool for arbitrary component analysis support.

The openness as a global system property is elaborated in remarkable work on Physics of Open Systems (POS) where the complexity of systems is perceived from systems dynamics (a complexity of movement) [3]. The described POS scientific methods and technologies enable the formation of scientifically proven systems ontological knowledge from its empirical descriptions embedded in a huge amount of semi-structured, multimodal, multidimensional, and heterogeneous data.

The main question arises: *Is it possible to upraise the SoS architecture as a reflection of successive and self-improve dimension reduction processes concerning basic POS principles applied on the persistent information resources base that preserves the SoS dynamics descriptions in the form of LDOs?*

Gaining a definitive answer to this question is far beyond the scope of this chapter. Its intended mission is to formulate a starting SoS-based framework model that may be further dynamically improved to better suit the static structure and dynamic behavior of complex SoS configurations. That is the reason why the reflexive integration of POS and different dimensional reduction methods, through an interoperability framework, have been proposed as the main contribution of this research chapter.

The rest of the chapter is organized as follows: Section 2—The background and motivation elaborate problem domain aspects of SoS multilevel multidimensional and representation, interoperability and integration aspects, architectural challenges, and framework-based approach with selected related work analysis and discussion. Section 3—System of Systems Hyper Framework Model (SoS-HFM)—states and elaborates the proposed model as a foundation for arbitrary component analysis method application. It is intended to serve as a reliable LDOs pool for incremental dimension reduction activities that converge to the SoS architecture upraise. Section 4—Conclusion is devoted to the concluding remarks, challenges, and the directions of future research activities. In the Appendices section, there are several Java code templates of SoS-HFM concepts presented. References—contains a list of references that have been analyzed and used as the problem domain origins and comparative research results evaluation basis.

2. Background and motivation aspects

Scientific research is a robust and dynamic practice that employs multiple methods toward investigating systems or phenomena including experimentation, description, comparison, and modeling. According to ref. [4], these methods, although often used in combination, appear more effective if used alone. Experimental methods are used to investigate the relationship(s) between two or more phenomena in a strictly controlled environment. Description methods utilize the observations and measurements of natural phenomena and their relationships, to collect the relevant data set that describes their behavior. Comparison is used to determine and quantify relationships between two or more phenomena by observing different groups that are, either by choice or circumstance, exposed to different treatments. Scientific knowledge cannot be obtained from empirical data by purely logical means because the ontologies of scientific and empirical knowledge differ significantly. Physics of Open Systems (POS), briefly introduced in the previous section [3], facilitate the generation of scientifically proven knowledge about the ontology of open systems via data mining techniques, applied on a huge amount of semi-structured, multimodal, and heterogeneous data that is dynamically generated throughout the SoS lifecycle. The identification of characteristic symmetries in an ontology model is used to simplify the structure and behavior of open systems over the state space defined by the ontology model.

The modeling is a well-established mechanism for struggle with the complexity that is the main obstacle of contemporary systems and solutions. The results of the modeling process are a single or a combination of physical and/or computer-based models of natural systems and/or phenomena that are afterward used as a framework for experiments and/or observations. Scientific development (progress) addresses the scientific approach to overall and sustainable development concerning the wide variety of contemporary problems that are either global or domain specific [5].

Concerning the SoS discipline, despite the inherent complexity, the large picture approach is usually the most promising one. The SoS exhibits an organized form of complexity and therefore cannot be accurately described by the traditional analysis techniques. The key concept of complexity science is universality, which is the idea that many systems in different domains exhibit phenomena with common underlying features that can be described using the same scientific models. Complexity science can provide a comprehensive, cross-disciplinary analytical approach that complements traditional scientific approaches that are focused on a specific observed subject in each domain. Complex systems are often characterized by many components that interact in multiple ways among each other and, potentially, with their environment too. These components form then dynamic networks of interactions, with vide variety of network topologies. They generally range from configurations with a small number of components that are involved in a large number of interactions, to configurations that involve the enormous number of components involved in a small number of interactions (**Figure 1**).

Interactions may generate novel information that makes it difficult to study components in isolation or to completely predict their future structure and/or behavior. The main challenge of complexity science is not only to have a sense of the parts and their connections but also to understand how these connections give rise to the whole. Advanced mathematical and computational modeling, analysis, and simulations are almost always required to investigate how these configurations are structured and change with time.

The growth of stakeholder-driven content has fueled a rapid increase in the volume and type of data that is generated, manipulated, analyzed, and archived. In addition, varied newer sets of sources, including sensors, Global Positioning Systems (GPS), automated trackers, and monitoring systems, are generating a huge amount of data multidimensional. These larger volumes of data sets, often termed *big data,* are imposing newer challenges and opportunities considering: storage, retrieval, analysis, visualization, and long-term archival. Computer-based analysis of massive data, emerging from complex systems structure and behavior, enables the recognition of embedded data/information/knowledge/wisdom (DIKW) patterns that contribute the further understanding of structure and behavior either of the wholes and/or its parts, thereby fostering the more accurate prediction of forthcoming structure and/or behavior. The second challenge raises directly from the multilevel and multidimensional nature of the artifacts that are consciously or unconsciously reflected through the complex systems' time and configuration-dependent state transitions. In **Figure 2**, a cognitive DIKW pyramid is presented that relates LDOs that may be generated with the inherent semantics in mind.

Different forms of data representation are well established and experienced in engineering practice. It is generally not the case with Information, Knowledge, and Wisdom, because they are dominantly context-dependent. The SoS persistency layer is a crucial component of SoS based analytics and possible drawbacks of LDO concepts need careful attention while specifying a supportive framework model. Although there is a huge amount of intellectual value embedded in arbitrary real-life systems

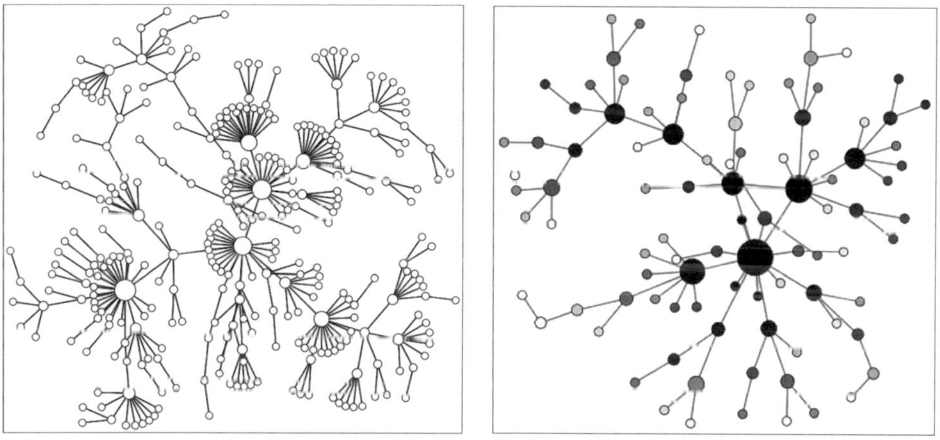

Figure 1.
Network representations of composite components [6, 7].

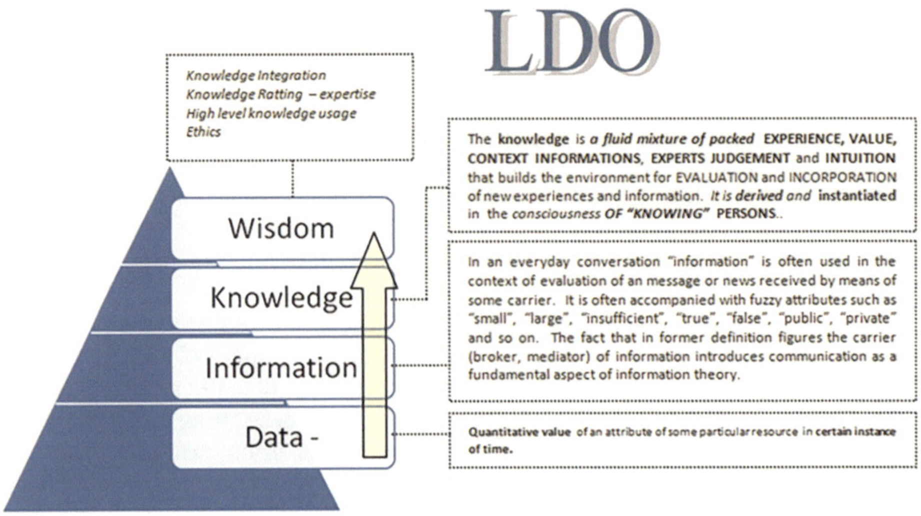

Figure 2.
The representation of LDOs genesis.

that are naturally represented in semi-structured or unstructured form, the contemporary legacy Enterprise Systems are still dominantly operated over structured repositories. Concerning the growing shift from SQL to NoSQL persistency the hybrid repository model seems an appropriate solution to start with.

In ref. [8], the authors elaborate Big Data management in the context of three inherent supportive dimensions: ***technology*** (dominantly related to storage, analytics, and visualization); ***people*** (addressing the human aspects); and ***processes*** (addressing technological and business approaches to management aspects). The semantic value, quality of data, and data security are stated as dominant challenging issues concerning the Big Data foundation of arbitrary SoS artifacts. The Framework-based approach to Big Data analytics application in high-level education environments is presented in ref. [9]. Although strictly conceptual the proposed framework model may be applied beyond the scope of the education domain. In ref. [10], there is the application of Linear Mixed Modeling (LMM) promoted as a flexible approach for scientific experimental data analysis. The nature of experimental data opens a challenging question of dataset quality metrics that may be proliferated to the SoS dynamic configurations instances. The multilevel ontological generation of semantic relations extracted from the significant amount of heterogeneous linguistic data, persisting in Big Data repositories, has been proposed by Popova et al. [11]. The solution is based on a specific XML format that enforces the interoperability of information across individual levels of generated multilevel ontology, for a particular problem domain. The software engineering perspectives of the Big Data foundation are surveyed in ref. [12]. The refinements of software development activities through the challenging aspects of corresponding Big Data concepts are discussed with the particular accent on architecture design, software quality insurance, and data quality assessment. In ref. [13] the intelligent systems design processes are discussed through multidimensional modeling of knowledge and knowledge transfer between internal components and external counterparts of arbitrary intelligent systems viewed as a four-dimensional (*grade*, *atomization*, *abstractness*, *timing*) cellular architecture. The engineering aspects of spatial data is an challenging domain for engineering disciplines that are based on different natural phenomena analysis and simulation like daylight illumination of residential buildings in ref. [14] where the representation of large data objects and its potential dimensionality reduction has an dramatic impact on urban planning.

Although computer-based analysis of massive data sets is in the field for a quite long time it is far from being a routine activity. From the architecture aspect, it is essential to separate the external repository from the internal dynamic storing and presentation layers and thereby hide the particular characteristics of persistent LDO form from its operational counterparts. This is the first pillar of the proposed SoS framework model.

The computational complexity of high-dimensional LDOs processing is the main obstacle for real-time or near real-time applications. The LDO's complexity reduction appears as a promising approach to the system/problem simplification process. Because less complex LDOs are easier to navigate, explore, visualize and analyze in different contexts they are more suitable for effective machine learning.

That is the main reason why several complexity reduction methods, based on different dimensionality reduction algorithms, have been proposed, formalized, applied, and verified. Among them, there are two main unsupervised methods worth mentioning: hierarchical clustering (HC) and principal component analysis (PCA). HC tries to build a tree-like structure with leaves representing the individual objects and nodes (pseudo objects) representing the clustering points of leaves with the highest degree of similarity. In further iterations, the individual clusters (as surrogates of clustered objects) replace the whole group and appears as individual objects with a certain accuracy payoff. PCA, on the other hand, creates a lower-dimensional representation of the initial data set on top of principal components as patterns encoding the highest variance in the data set, trying to preserve as much as possible of the original data set variance in process of dimensionality reduction.

Reducing the complexity of a particular LDO has its payoff in the accuracy of the reduced counterpart. The quality of a dimensionality reduction method is measured by its ability to gain the lowest possible complexity with the highest possible accuracy. Being unsupervised, HC and PCA methods are better suited for the generation of sustainable simpler LDOs, and consequently simpler SoSs configurations, rather than their verification.

Due to the fundamental focus of this chapter, the rest of the section is devoted to a more detailed elaboration of solely the PCA methods-related publications analysis. The mathematical elaboration is completely avoided due to the huge amount of references that have excessively addressed the foundation. A remarkable complete, simplified, step-by-step analysis of the original PCA method is presented in ref. [15], through five consecutive steps that lead to the data set dimensionality reduction. It starts with the standardization of the initial variable (dimension) range to comparable scale, to eliminate the possible supremacy of dominant instances, followed by the calculation of the covariance matrix of all possible pairs of scaled variables (dimensions) to uncover the correlation nature of each possible variable pairs. In the third step, the principal components isolation is performed by computing the eigenvectors of the covariance matrix and ordering them by their eigenvalues in descending order. This process isolates the principal components, which are the surrogates of correlated dimensions, in order of their significance. In the fourth step, the Feature vector is created by selecting the representative subset of principal components that leads to the desired dimensional reduction with preferable accuracy. The last step is the generation of reduced dimensional data set by data recasting over selected principal components. The mathematical foundation of linear PCA is gradually presented in ref. [16], and joined with the context of the previously referenced article completes its light-weighted approach. In ref. [17] the original row-based two-dimensional principal component analysis (2DPCA) and its extensions in the 2D image processing domain, have been discussed. The overview of several extension frameworks have been presented (bilateral projection nonlinear and iterative, kernel-based, supervised (with four variations), alignment-based, and random (with three variations). The robustness of all of the elaborated extensions has exhibited a performance increase in comparison to the original 2DPCA in image recognition, which has been a traditional application domain of PCA methods. PCA is discussed in ref. [18] through the extensive survey of five RPCA published models, with their comparative analysis in the context of video stream background management. In ref. [19] the modification of classical PCA (MPCA) with subspace learning framework based on multiple similarity measurements. In ref. [20] a comprehensive review and future PCA development have been presented. Although the reference is almost six years old, it has been used for the sake of the overall problem domain clarification where PCA is addressed as the linear exploratory tool for data analysis. Due to its application in different fields, the initial PCA has been modified in several ways to better suit the domain-specific characteristics. The authors enlist and discuss: static and dynamic functional PCA (FPCA), simplified PCA (SPCA), robust PCA (RPCA), and symbolic data PCA (SDPCA). In ref. [21], authors present the modification of nonlinear Kernel PCA (KPCA), with adaptive feature—Adaptive Kernel PCA (AKPCA) that is integrated with the gray relation analysis (GRA) for fault detection in complex nonlinear chemical processes.

The comprehensive analysis of different domain-specific PCA applications is far beyond the reasonable research effort. In ref. [22] the application of state-space functional PCA (SS-FPCA), as a 3-level hierarchical model built under the state-space model framework, for identification of spatiotemporal patterns based on satellite remote sensing of lake water quality in the form of time series of spatial images with missing observations. The authors of ref. [23] elaborate the Principal Component-based support vector machine (PC-SVM) as a hybrid machine learning technique that combines PCA and SVM to cope with the potential software defects especially concerning the mission-critical software systems. In ref. [24] the contemporary challenging study of the implementation of machine learning methods for identification of patients affected by COVID-19 based on X-ray images. Two commonly used classifiers were selected: logistic regression (LR) and convolution neural networks (CNN) joined with PCA for complexity reduction and shorting the elapsed time to gain the quality diagnostic answer. The complex boundary generation method, presented in ref. [25], illustrates an practical application of dimensionality variation through the recursive search of the optimal residential building outer shape form, based on variable set of parameters.

This short survey and the much broader repertoire of similar research articles fully qualify the research motivation of this chapter, the formulation of supportive SoS Hyper Framework Model.

Physics of Open Systems is the additional paradigm for SoS Hyper Framework Model development where the system is considered as a tool where the knowledge and sense of its complexity are harvested. It is necessary to reference [3] for the rest of the relevant influencers. The directly influencing POS intellectual machine analytical core technologies that support systems: reconstruction through ontology mode variations; examination based on communication model variations; design based on state model variation; empirical context formation—the generation of LDOs in this chapter context; solutions behavior generation—the dynamic representation of varied model;

visualization. Its formulation is transformed in the formulation of POS dimensions of the system model in the context of this chapters' SoS Hyper Framework Model proposal.

Software-supported frameworks of the arbitrary kind are usually targeted in model-driven software development (MDSD) and model-based system engineering (MBSE) approaches. They are closely related to the general architecture modeling paradigms and constitute the core of different contemporary enterprise architecture (EA) frameworks. There are several EA Frameworks proposed and specified, among which the most advocated are:

- *Zachman Framework* (ZF), the framework for enterprise architecture (EA). According to ref. [26], it is an EA ontology-based on the visualization of an enterprise and its inherent information system components and their relations from different perspectives. It is a two-dimensional classification scheme structured as a matrix containing 36 cells, each of them focusing on one dimension or perspective of the enterprise. Although popular in EA academic education and learning environments, due to its inexplicable practical utility, it has not reached substantial practical achievements and is expected to fade out from the EA community in near future [27].

- *The Open Group Architecture Framework* (TOGAF), is declared as the most commonly used enterprise architecture framework [28]. Due to its architecture development method (ADM) orientation, it is considered dominantly process-oriented. The overall TOGAF process is organized into four groups: *Business architecture*; *Application architecture*; *Data architecture*; and *Technical architecture*. Due to its OO community origins, it has been announced as a promising EA framework. On the contrary, the extensive search through the contemporary available resources on Global Network shows that there is still a lack of usable TOGAF software support tools which makes its practical usability questionable.

- *Federal Enterprise Architecture Framework* (FEAF), according to ref. [29], has offered a shared approach for the consolidation of strategic, business, and technology management as a component of organization design and performance management. It is composed of 6 interconnected Reference Models (Strategy; Business; Data; Applications; Infrastructure; and Security), each relating to a sub-architectural domain of the framework that is linked together through the Consolidated Reference Model (CRM). The main problem FEAF faces today is the lack of real verifiable success stories even by the enterprise architects working for the U.S. Government [27].

- *CMMI Framework* – Capability Maturity Model Integrated, Software Engineering Institute (SEI) Framework. Although originally tailored toward software, the latest CMMA Framework version is more general and applies to hardware, software, and service development across all industries [30]. It is a process-oriented framework, whose main purpose is to assess the maturity of an organization's processes and to provide guidance on their improvement to deliver high-quality products. In the CMMI model, version 1.3, there are 22 process areas defined, together with the process-related goals and the set of activities that are often used to meet them. Because the CMMI Framework is based on well-defined ontology, classification system, experience, training, and appraisal infrastructure (governed by SEI), it is expected to evolve into a de-facto standard for the maturity classification of at least Software Development Companies.

- *International Council On Systems Engineering* (INCOSE)—System of Systems Framework [31]- is closely related to the Model-Based Systems Engineering (MBSE) approach that emerged from the INCOSE projects in the SoS engineering domain. It is clustered over three main concepts: ***model*** (*a formally simplified version of an entity of interest*), ***systems thinking*** (*a holistic approach to interacting entities and their components*) [32], and ***systems engineering*** (*transdisciplinary and integrative approach to the engineering of interacting entities, based on systems principles and concepts in the context of scientific, technological and management methods application, through the entire life cycle*). MBSE does not strictly prescribe any process framework the arbitrarily selected process model has to address the four essential systems engineering domains: requirements/capabilities; the static structure (systems architecture or topology) [33]; the dynamic structure (systems behavior); verification and validation aspects [34] (is it the right system? and if is, is the system right?). In ref. [35] there is a remarkable well-illustrated approach to SoS mission needs break down to capabilities and functions through the architecture framework and related ontology that has inspired several concepts of the SOS Hyper Framework Model specification.

The related work analysis shows that there is a tremendously large number of documents, studies, standards, procedures, and scientific articles that dominantly address particular aspects of the Principal Component Analysis approach to handle the dimensionality reduction problem, but fare fewer references concerning POS paradigm and its implementation aspects in the context of SoS.

On the other hand, there is also a lack of research concerning the interoperability framework approach with the integration mission. These facts favor the large-picture-based approach facilitating the Generic System of Systems Framework that sustainably orchestrates: Domain-Specific and Generic concepts and dimensions of complex SoS configurations that are opened for arbitrary POS and PCA methods extension. The System of Systems Hyper Framework Model, presented in the next section of this chapter, is considered as a first step toward the established goal. The collaborative frameworks, like one elaborated in ref. [36], has served as an initial framework specification of the proposed model presented in this article.

3. The foundation for open component analysis: SoS hyper framework model (SoS-HFM)

3.1 Why framework-based approach?

In general, a framework may be defined as a real or conceptual foundation, with a specified level of complexity that serves as a support or a guide for the building of a particular artifact or performing a particular activity by expanding and specializing the generic structure that specifies the family of interrelated products and/or procedures. Framework favors reusability by managing the overall control flow and orchestration of dynamically configured components in an inversion of control way. There are two major categories of contemporary framework: non-software empowered (usually represented as a set of structured and/or semi-structured documents) and software empowered (software supported collaborative/cooperative environments supporting the digital transformation of problem domain). The development of software empowered interoperability frameworks have usually been preceded by the intensive and time-consuming: specification; modeling; and meta-modeling activities performed and managed within the scope of related projects and/or portfolios.

3.2 Why interoperability-based approach?

In the context of this chapter, cooperation and collaboration of independently developed or specified systems may be generally achieved through total homogenization; harmonization; adaptation; and orchestration interoperability principles. The total homogenization principle states that all of the related systems have to be designed or redesigned to achieve absolute compliance with the globally accepted and standardized template. Homogenization is the most radical, heavy-weight, approach. A harmonization principle is a lightweight approach that assumes the definition and standardization of interfaces that hide the internal characteristic of participating systems and enable their cooperation/collaboration over the unique communication protocol and through the standardized interfaces only. The participating systems need to be functionally complete while retaining the structural diversity. The adaptation interoperability principle assumes cooperation/collaboration of functionally and structurally incomplete systems where participating systems are homogenized up to the functional and structural completeness, in a virtual or real way (where homogenization payoff may substantially differ from system to system) and harmonized afterward. Adaptation interoperability may be seen as a middle-weight approach. It is more complex than the harmonization but, compared to the total homogenization, significantly more acceptable and achievable faster. The orchestration interoperability is the ability of the heterogeneous systems, with arbitrary functionality and topology, to interact toward the mutually beneficial dynamically configured mission, build through the functional and/or structural orchestration of participating systems features and/or resources. Each system retains and shares everything it can perform and/or deliver and delegates and/or acquires everything that is beyond its scope, but available as a mutual benefit of a current configuration. The orchestration interoperability is an example of broker-based service-oriented dynamic architecture that may be formally described by the swarm intelligence concepts.

3.3 Why the interoperability framework approach?

If carefully combined the best characteristics of two, previously discussed promising concepts, may result in an empowered solution capable of handling the growing complexity of SoS dynamic configurations associated with the multidimensional and multilevel LDOs. The development of interoperability framework generally requires a multi stakeholder process and the long-term vision of a highly reusable generic solution that, in the context of SoS-HFM, may impact the overall ontology, configurable topology, state-driven behavior, and time and context dependent LDOs: creation, processing, storing, retrieval and visualizing.

The starting point of SoS and POS paradigms integration is the formulation of a hybrid system meta-model that combines multidimensionality and context-based multilevel features. The meta-concept (MetaConcept) of SoS-HFM is modeled as a typed composite MetaElement presented in **Figure 3**.

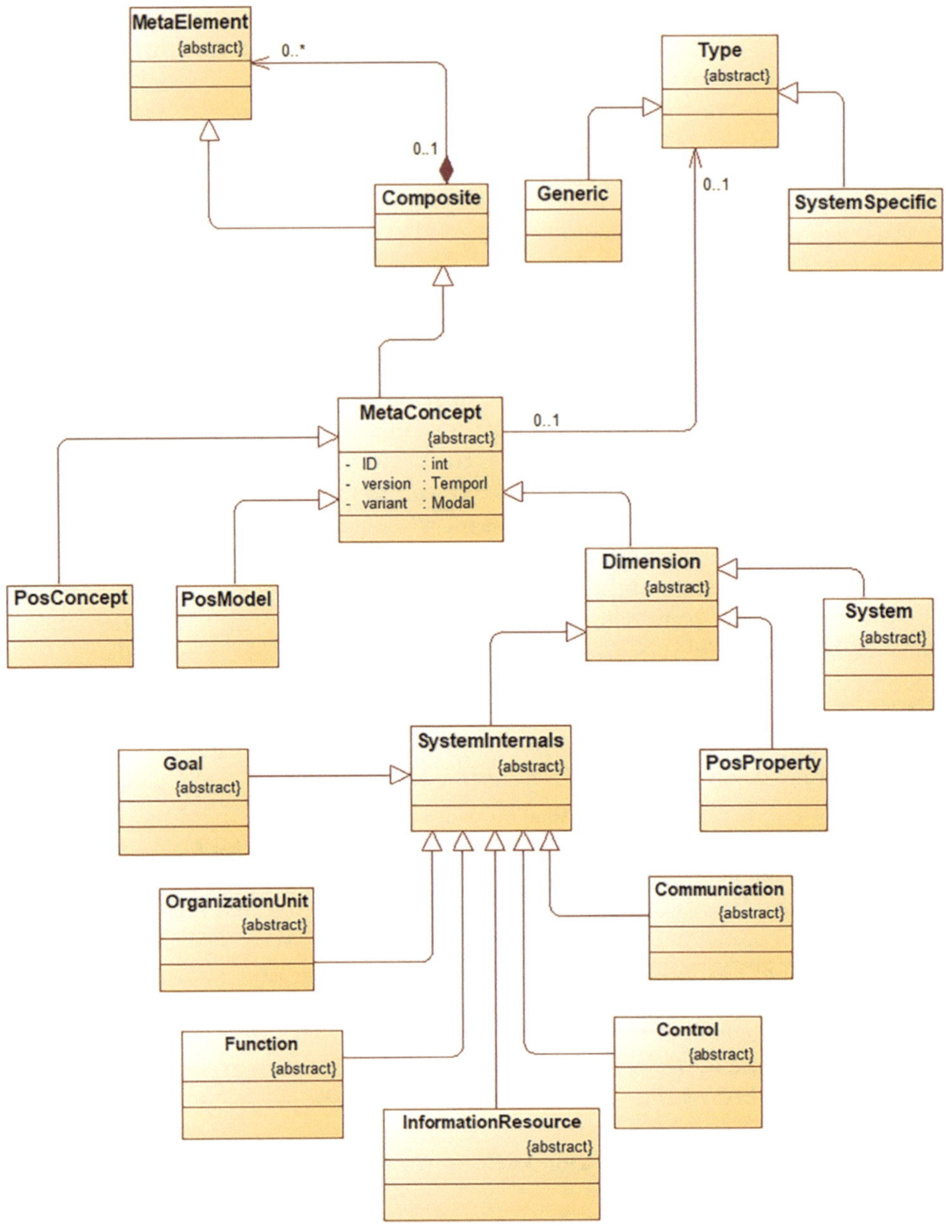

Figure 3.
SoS-HFM—meta concept model (a part of).

The composition property enables arbitrary topology creation in different contexts (determined by variant property as *Modal* object, and version property as *Temporal* object), and type associated property that enables the classification over an open set of classifiers with currently two specified: Generic and system specific.

The open set of MetaConcept specializations is currently composed of three elements: PosConcept (relates the framework ontology with POS System ontology); PosModel (relates the framework ontology with POS Model ontology segment); and

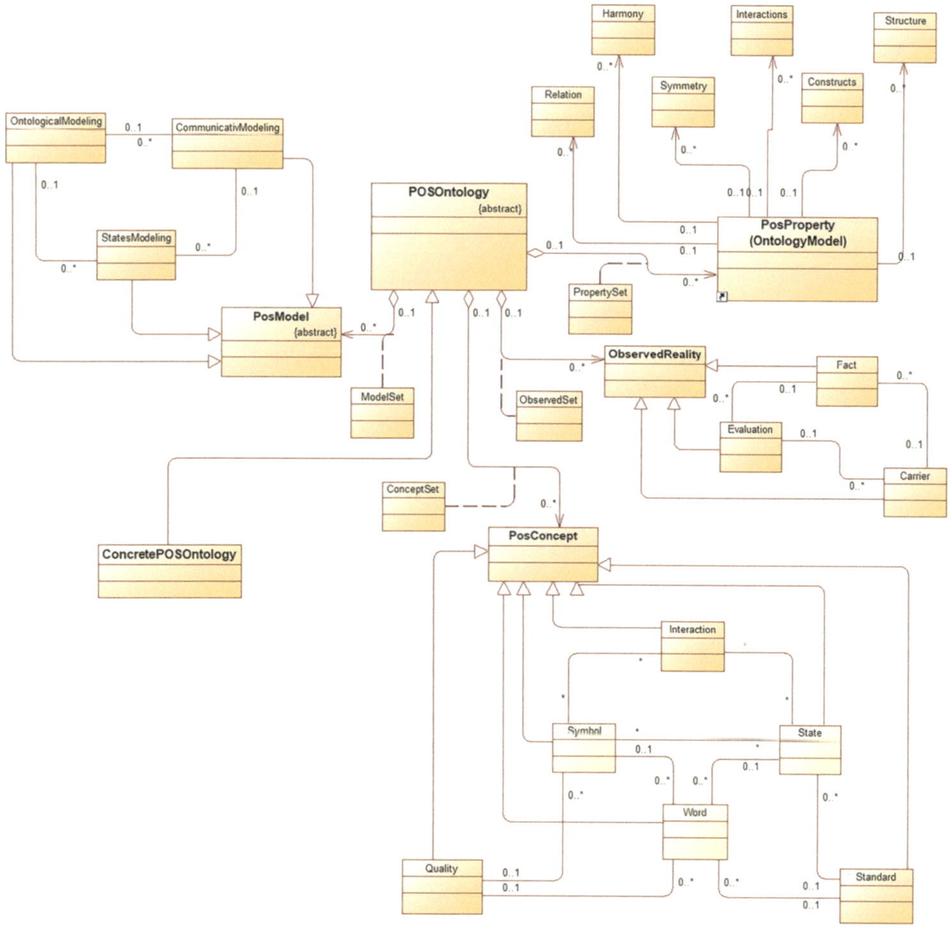

Figure 4.
The POSOntology model.

Dimension (abstract MetaConcept that clusters the fundamental SoS-HFM object of interest). The Dimension specialization is further specialized by an open set currently 3 concepts: System and POSProperty that will be defined later on and System Internals with fundamental internal dimensions of an arbitrary general system (Goal; OrganizationUnit; Function; InformationResource; Control; and Communication) that are elaborated in following segments of this Section.

The POSOntology model, developed in compliance with [3], is presented in **Figure 4**. POSOntology is an LDO that is composed of four collections:

1. PosProperty collection, organized by the PropertySet associative class, is an LDO that is configured from six predefined POS contextual dimensions (Relation; Harmony; Symmetry; Interactions; Constructs; and Structure),

2. PosModel collection, organized by the ModelSet associative class, represents the POS model triangle composed of contextual modeling dimensions (OntologycalModeling; CommunicativModeling; and StatesModeling);

3. ObservedReality collection, organized by the ObservedSet associative class, represents the POS model triangle composed of contextual observed dimensions (Fact, Evaluation, and Carrier);

4. PosConcept collection, organized by the ConceptSet associative class, represents two interrelated POS triangles composed of contextual concept dimensions (Symbol; State; Word) and (Interaction; Quality; Standard)

ConcretePOSOntology is a domain or system-specific specialization of POSOntology that forms the ontology segment of SoS-HFM and is the targeted ontology of arbitrary orchestrated POS methods.

The detailed System Dimension of the SoS-HFM ontology meta-model is presented in **Figure 5**. The system is an LDO, defined as the SoS-HFM dimension that represents the organized set of interrelated components that are orchestrated (configured) with the specific mission in mind.

It is composed of five embedded associative collections and one inherited associative collection. The five embedded associative collections are:

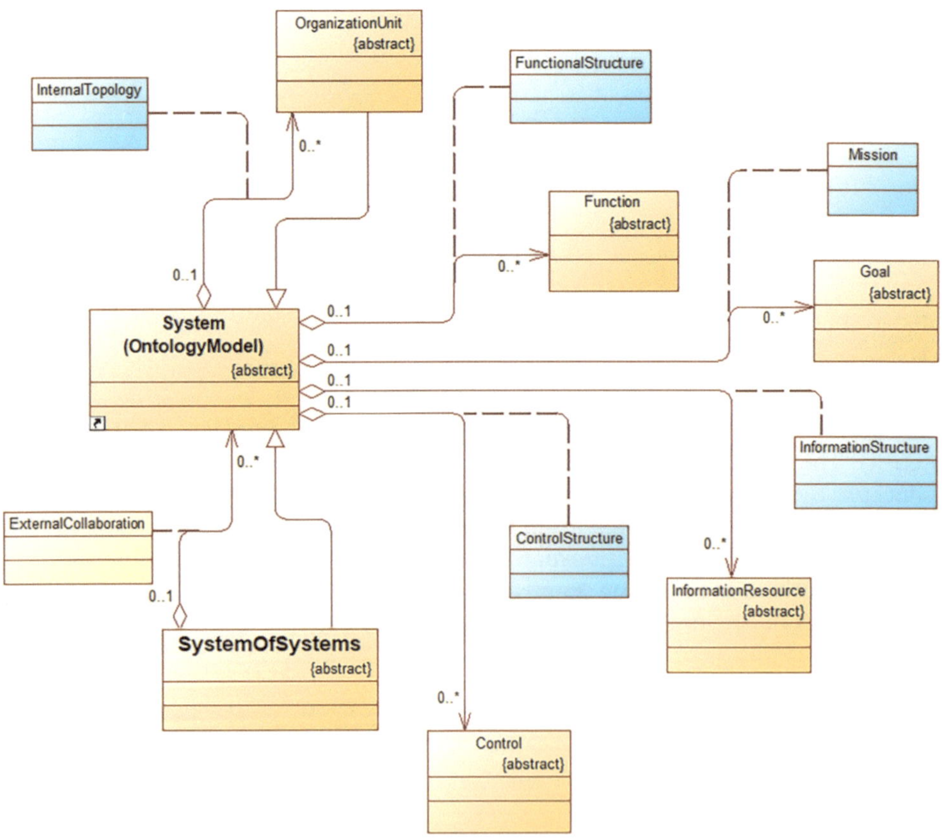

Figure 5.
The system dimension of SoS-HFM meta model.

1. Goal set collection, organized by the Mission associative class, that enables the modeling of domain-specific systems mission (the goal or set of goals that justifies the existence of a system);

2. OrganizationUnit set collection, organized by the InternalTopology associative class, that forms the instances of internal systems architecture;

3. Function set collection, organized by the FunctionalStructure associative class, that forms thy configurations of internal activities that are spread over the OrganizationUnit topology in favor of the overall Mission;

4. InformationResource collection, organized by the InformationStructure associative class, that forms the supporting information infrastructure of a Functional topology;

5. Control set collection, organized by the ControlStructure associative class, forms the monitoring, control, and management of the configuration instances in favor of gaining the overall systems Mission.

The inherited associative collection, organized by the ExternalCollaboration associative class, represents the SoS dimension of the SoS-HFM meta-model, with SoS defined as a specialization of system meta concept.

The SoS-HFM dimensionality reduction meta-model is presented in **Figure 6**. The key meta concepts defined are the configuration and the LargeDataObject.

1. Configuration is the specialization of SoS meta-class with three associative collections: Orchestrated method represents the aggregation of orchestrated dimensionality reduction methods for the particular configuration instance, managed by the dimension meta-class; ConcretePOSOntology relates configuration with the set of associated POS ontologies; and LargeDataObject composite collection that is either original or dimensionally reduced counterpart.

2. LargeDataObject (LDO) encapsulates the dimensionality information and the status of each dimension that is either reducible (Reducible) or not, for an unreduced instance, and reduced (Reduced) if reducible in a reduced instance. LDO is associated with a particular ontology (ConcretePOSOntology). Being a MetaConcept LDO inherits temporal and modal characteristics and marks the particular instances with.

In **Figure 7**, the conceptual model of SoS-Hyper Framework software architecture is presented. It is modeled as MVC architectural pattern where: SoSHFModel—handles the framework dynamic data structure; SoSHFView—encapsulates the collection of SoSHFModel visualization methods, and SoSHFControler—supports framework dynamics. ServiceControler handles an open set of framework services with currently specified: DataSetBuilder; SesionManager; ActivityTracker; and SecurityManager. RepositoryManager encapsulates an open set of repository handlers with currently specified: RelationalRepository; and NoSQL open set of Non-relational repository handlers.

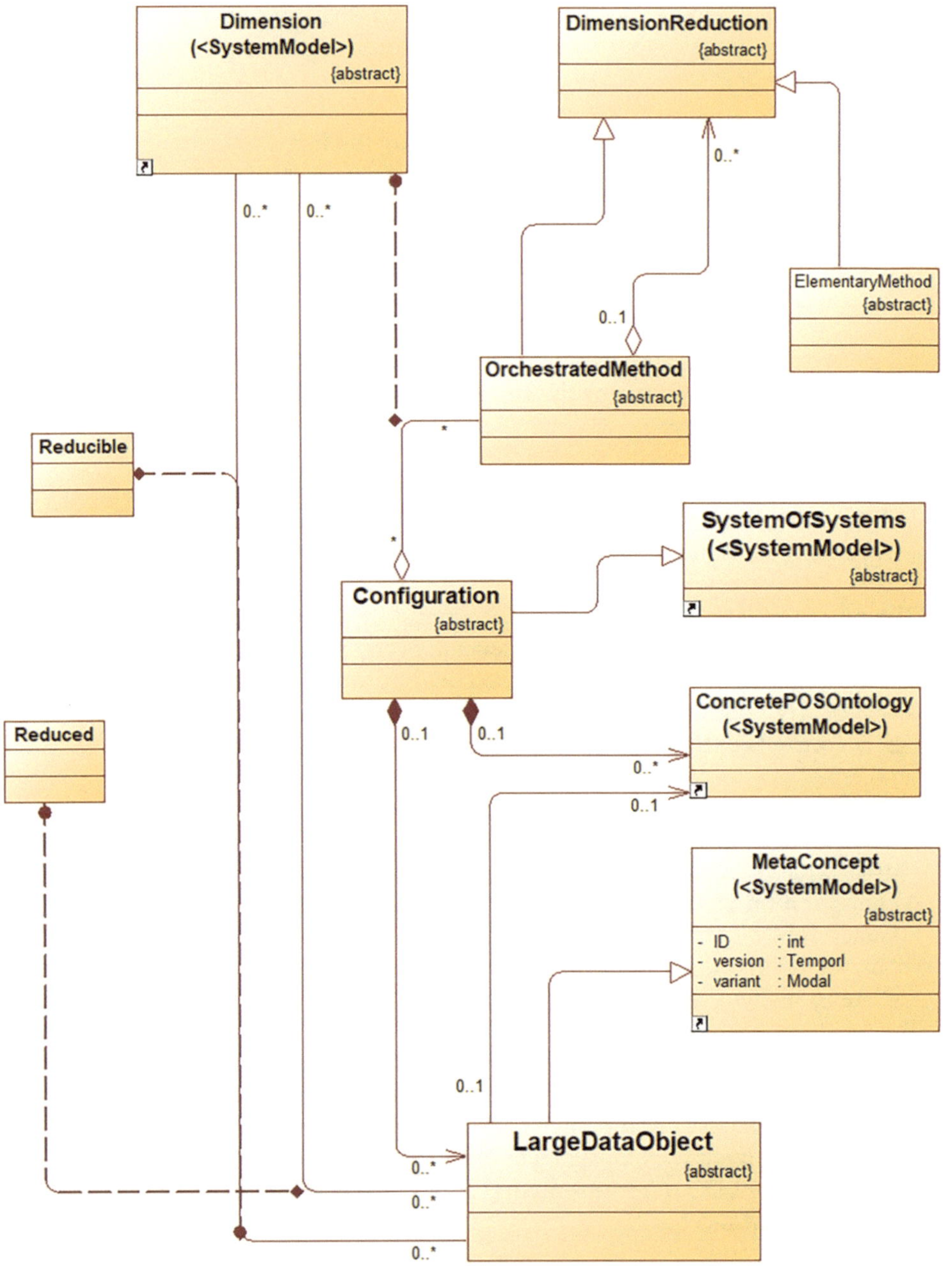

Figure 6.
The SoS-HFM dimensionality reduction meta model.

In **Figure 8**, there is a more detailed conceptual model of the SoSHF Controller segment that encapsulates POS and dimensionality reduction methods (PCA) presented. It utilizes a two-component interface: *AccessIR* (manipulation of arbitrary information resources designated as LDOs in this chapter); and *DimensionAccess* (manipulation of the collection of dimensions embedded in LDOs).

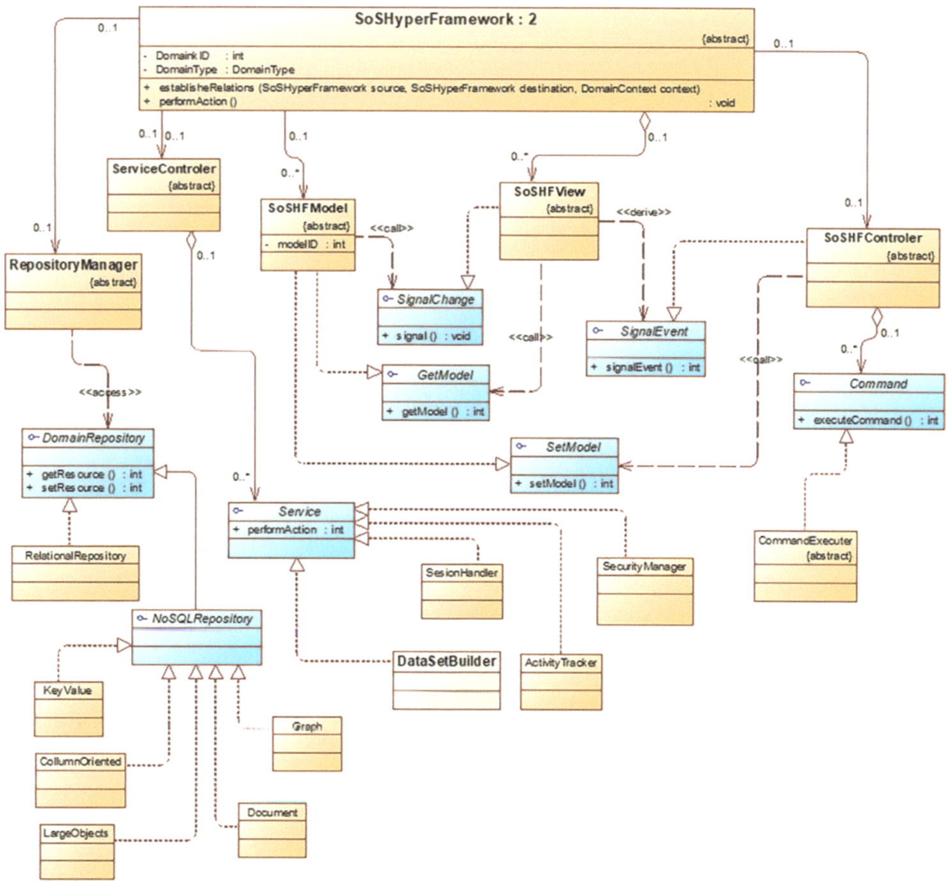

Figure 7.
SoS-hyper framework conceptual model (MVC).

3.4 The discussion

As previously mentioned, the main challenge of complexity science is not only to understand and manage the individual components and their connections that make the configuration (static structure) but also to understand how these connections affect the whole. The Foundation for Open Component Analysis is specified on top of three synergic contexts, The Physics of Open Systems ontology, System specification as an SoS composition with multidimensional and multilevel instantiation capabilities, and the simplification mechanisms based on the extendible set of the Principal Component Analysis methods. The orchestration is specified through the System of Systems Hyper-Framework Model where advanced mathematical and computational modeling, analysis, and simulations may be applied to investigate how these orchestrated configurations are structured and change with time.

4. Conclusion

Considering the inherent complexity of System of Systems, the mission of creating the foundation of Opened Component Analysis emerged with an SoS- Hyper

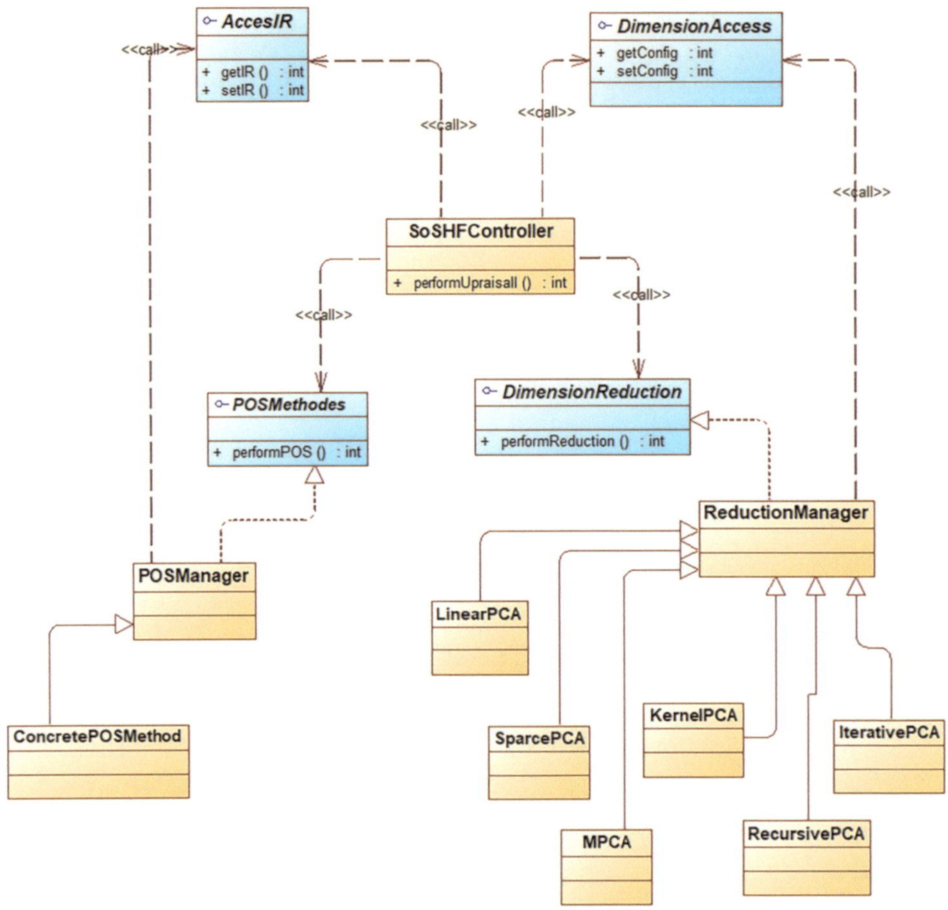

Figure 8.
SoSHFController—POC and PCA dimensionality reduction.

Framework Model that has inherited the challenging aspects of Physical Open System fundamental concepts and methods and dimensionality reduction methods by the orchestration of an opened set of Principle Component Analysis methods. The main characteristics of the proposed framework model are elaborated in the Introductory section, founded through the Background and motivation section with related work analysis, and explicated in Section 3 with the presentation of the key aspects of SoS-HFM. In Appendices A.1, there is a sample of Java code generated from the corresponding Configuration meta class presented.

The proposed model is attended to serve as a starting specification for the development of the future: open, heterogeneous, cooperative/collaborative, a service-oriented software framework that may be tailored according to the SoS configurations. These are also the main directions of future research and work.

Conflict of interest

The authors declare no conflict of interest.

Appendices

A.1 SoSHyperFramework: model-based generated Java code skeleton

```
/*****************************************************************
* Module: SoSHyperFramework.java
* Author: Branko
* Purpose: Defines the Class SoSHyperFramework
*****************************************************************/
import java.util.*;
/** @pdOid 5199ec80-82c8-4224-adf4-6f79e037a32c */
public abstract class SoSHyperFramework {
   /** @pdOid 29f37534-7bc3-46d9-9fa5-f8996a25eb8d */
   private int domainkID;
   /** @pdOid 3f9f6a74-0c4c-448d-878e-1e4d6b216519 */
   private DomainType domainType;
   public java.util.Collection frameworkRelationsB;
   /** @pdRoleInfo migr=no name=DomainContext assc=association3 coll=java.
   util.Collection impl=java.util.HashSet mult=* */
   public java.util.Collection<DomainContext> domainContext;
   /** @pdRoleInfo migr=no name=SoSHFControler assc=association6 mult=0..1 */
   public SoSHFControler soSHFControler;
   /** @pdRoleInfo migr=no name=SoSHFView assc=association7 coll=java.util.
       Collection impl=java.util.HashSet mult=0..* type=Aggregation */
   public java.util.Collection<SoSHFView> soSHFView;
   /** @pdRoleInfo migr=no name=SoSHFModel assc=association8 coll=java.util.
       Collection impl=java.util.HashSet mult=0..* */
   public java.util.Collection<SoSHFModel> soSHFModel;
   /** @pdRoleInfo migr=no name=RepositoryManager assc=association9 mult=0..1 */
   public RepositoryManager repositoryManager;
   /** @pdRoleInfo migr=no name=ServiceControler assc=association10 mult=0..1 */
   public ServiceControler serviceControler;
   public CooperatingDomains[] frameworkRelationsA;
   /** @param source
   * @param destination
   * @param context
   * @pdOid 744debe3-5548-4295-a6ea-328715b03caa */
   public SoSHyperFramework establisheRelations(SoSHyperFramework source,
SoSHyperFramework destination, DomainContext contcxt) {
      // TODO: implement
   }
   /** @pdOid 75f8bfff-a41b-47be-a366-c6b44131030d */
   public void performAction() {
      // TODO: implement
   }
   /** @pdGenerated default getter */
   public java.util.Collection<DomainContext> getDomainContext() {
      if (domainContext == null)
         domainContext = new java.util.HashSet<DomainContext>();
```

```
        return domainContext;
    }
    /** @pdGenerated default iterator getter */
    public java.util.Iterator getIteratorDomainContext() {
        if (domainContext == null)
            domainContext = new java.util.HashSet<DomainContext>();
        return domainContext.iterator();
    }
    /** @pdGenerated default setter
    * @param newDomainContext */
    public void setDomainContext(java.util.Collection<DomainContext>
newDomainContext) {
    removeAllDomainContext();
    for (java.util.Iterator iter = newDomainContext.iterator(); iter.hasNext();)
        addDomainContext((DomainContext)iter.next());
    }
    /** @pdGenerated default add
    * @param newDomainContext */
    public void addDomainContext(DomainContext newDomainContext) {
        if (newDomainContext == null)
            return;
        if (this.domainContext == null)
            this.domainContext = new java.util.HashSet<DomainContext>();
        if (!this.domainContext.contains(newDomainContext))
            this.domainContext.add(newDomainContext);
    }
    /** @pdGenerated default remove
    * @param oldDomainContext */
    public void removeDomainContext(DomainContext oldDomainContext) {
        if (oldDomainContext == null)
            return;
        if (this.domainContext != null)
        if (this.domainContext.contains(oldDomainContext))
            this.domainContext.remove(oldDomainContext);
    }
    /** @pdGenerated default removeAll */
    public void removeAllDomainContext() {
        if (domainContext != null)
            domainContext.clear();
    }
    /** @pdGenerated default getter */
    public java.util.Collection<SoSHFView> getSoSHFView() {
        if (soSHFView == null)
            soSHFView = new java.util.HashSet<SoSHFView>();
        return soSHFView;
    }
    /** @pdGenerated default iterator getter */
    public java.util.Iterator getIteratorSoSHFView() {
        if (soSHFView == null)
            soSHFView = new java.util.HashSet<SoSHFView>();
```

```
      return soSHFView.iterator();
   }
   /** @pdGenerated default setter
   * @param newSoSHFView */
   public void setSoSHFView(java.util.Collection<SoSHFView> newSoSHFView) {
      removeAllSoSHFView();
      for (java.util.Iterator iter = newSoSHFView.iterator(); iter.hasNext();)
         addSoSHFView((SoSHFView)iter.next());
   }
   /** @pdGenerated default add
   * @param newSoSHFView */
   public void addSoSHFView(SoSHFView newSoSHFView) {
      if (newSoSHFView == null)
         return;
      if (this.soSHFView == null)
         this.soSHFView = new java.util.HashSet<SoSHFView>();
      if (!this.soSHFView.contains(newSoSHFView))
         this.soSHFView.add(newSoSHFView);
   }
   /** @pdGenerated default remove
   * @param oldSoSHFView */
   public void removeSoSHFView(SoSHFView oldSoSHFView) {
      if (oldSoSHFView == null)
         return;
      if (this.soSHFView != null)
      if (this.soSHFView.contains(oldSoSHFView))
         this.soSHFView.remove(oldSoSHFView);
   }
   /** @pdGenerated default removeAll */
   public void removeAllSoSHFView() {
      if (soSHFView != null)
         soSHFView.clear();
   }
   /** @pdGenerated default getter */
   public java.util.Collection<SoSHFModel> getSoSHFModel() {
      if (soSHFModel == null)
         soSHFModel = new java.util.HashSet<SoSHFModel>();
      return soSHFModel;
   }
   /** @pdGenerated default iterator getter */
   public java.util.Iterator getIteratorSoSHFModel() {
      if (soSHFModel == null)
         soSHFModel = new java.util.HashSet<SoSHFModel>();
      return soSHFModel.iterator();
   }
   /** @pdGenerated default setter
   * @param newSoSHFModel */
   public void setSoSHFModel(java.util.Collection<SoSHFModel>
newSoSHFModel) {
      removeAllSoSHFModel();
```

```
        for (java.util.Iterator iter = newSoSHFModel.iterator(); iter.hasNext();)
            addSoSHFModel((SoSHFModel)iter.next());
    }
    /** @pdGenerated default add
    * @param newSoSHFModel */
    public void addSoSHFModel(SoSHFModel newSoSHFModel) {
        if (newSoSHFModel == null)
            return;
        if (this.soSHFModel == null)
            this.soSHFModel = new java.util.HashSet<SoSHFModel>();
        if (!this.soSHFModel.contains(newSoSHFModel))
            this.soSHFModel.add(newSoSHFModel);
    }
    /** @pdGenerated default remove
    * @param oldSoSHFModel */
    public void removeSoSHFModel(SoSHFModel oldSoSHFModel) {
        if (oldSoSHFModel == null)
            return;
        if (this.soSHFModel != null)
        if (this.soSHFModel.contains(oldSoSHFModel))
            this.soSHFModel.remove(oldSoSHFModel);
    }
    /** @pdGenerated default removeAll */
    public void removeAllSoSHFModel() {
        if (soSHFModel != null)
            soSHFModel.clear();
    }
}
```

A.2 A nomenclature list

ADM	Architecture Development Method
DIKW	Data/Information/Knowledge/Wisdom pattern
HFM	Hyper-Framework Model
LDO	Large Data Object
LMM	Linear Mixed Modeling
MBSE	Model Based Systems Engineering
MDSD	Model-Driven Software Development
NoSQL	Not only Structured Query Language
PCA	The Principal Component Analysis
POS	The Physics of \open \systems
SoS	The System of Systems
SQL	Structured Query Language
XML	Extensible Markup Language

Author details

Ana Perišić[1] and Branko Perišić[2]*

1 Faculty of Technical Sciences, University of Novi Sad, Novi Sad, Serbia

2 University Singidunum Belgrade, Belgrade, Serbia

*Address all correspondence to: bperisic@singidunum.ac.rs

References

[1] Gartner Glossary. Available from: https://www.gartner.com [Accessed: January 30, 2022]

[2] Guide to the Systems Engineering Body of Knowledge (SEBoK), version 2.5. Available from: https://www.sebokwiki.org/ [Accessed January 30, 2022]

[3] Kachanova TL, Fomin BF, Fomin OB. Generating scientifically proven knowledge about ontology of open systems: Multidimensional knowledge-centric system analytics. In: Ontology in Information Science. London: IntechOpen; 2017. DOI: 10.5772/intechopen.72046

[4] Carpi A, Anne EE. The Practice of Science: An Introduction to Research Methods. available from: https://www.visionlearning.com/en/library/Process-of-Science/49/The-Practice-of-Science/148 [Accessed: January 30, 2022]

[5] Schneegans S, Straza T, Lewis J, editors. UNESCO Science Report: The Race Against Time for Smarter Development. Paris: UNESCO Publishing; 2021

[6] Complexity. Available from: https://www.complexity-explorables.org/explorables/clustershuck/ [Accessed January 10, 2022]

[7] Complexity. https://www.complexity-explorables.org/explorables/knitworks/ [Accessed January 1, 2022]

[8] Rossi R, Hirama K. Characterizing big data management. Issues in Informing Science and Information Technology. 2015;**12**:165-180

[9] Klašnja-Milićević A, Ivanović M, Budimac Z. Data science in education: Big data and learning analytics. Computer Applications in Engineering Education. 2017;**25**(6):1066-1078. DOI: 10.1002/cae.21844

[10] Ćwiek-Kupczyńska H, Filipiak K, Markiewicz A, et al. Semantic concept schema of the linear mixed model of experimental observations. Science Data. 2020;**7**:70. DOI: 10.1038/s41597-020-0409-7

[11] Maryna Popova, Larysa Globa and Rina Novogrudska. Multilevel ontologies for big data analysis and processing, Proceedings of the 9th International Conference on Applied Innovations in IT, (ICAIIT), 2021. Koethen (Germany), 28 April 2021

[12] Davoudian A, Liu M. Big data systems: A software engineering perspective. ACM Computing Surveys. 2020;**53**:5

[13] Konstantin Kostenko .2020. Knowledge flows processes at multidimensional intelligent systems, Russian Conference on Artificial intelligence (RCAI 2020), October, 10–16, 2020, Moscow, Russia

[14] Perišić A, Lazić M, Obradović R, et al. Daylight and urban morphology: A model for analysing the average annual illumination of residential housing. Tehnički vjesnik. 2016;**23**(5):1343-1350. DOI: 10.17559/TV-20150526191843

[15] Jaadi Z. A Step-by-Step Explanation of Principal Component Analysis (PCA). Available from: https://builtin.com/data-science/step-step-explanation-principal-component-analysis [Accessed: January 20, 2022]

[16] Keho Y. The basics of linear principal components analysis. In: Sanguansat P,

editor. Principal Component Analysis. London: InTech; 2012. Available from: http://www.intechopen.com/books/principal-component-analysis/the-basic s-of-principal-component-analysis

[17] Sanguansat P. Two-dimensional principal component analysis and its extensions. In: Sanguansat P, editor. Principal Component Analysis. London: InTech; 2012. Available from: http://www.intechopen.com/books/principal-component-analysis/2dpca-and-its-extensions

[18] Guyon C, Bouwmans T, Zahzah E-h. Robust principal component analysis for background subtraction: Systematic evaluation and comparative analysis. In: Sanguansat P, editor. Principal Component Analysis. London: InTech; 2014. Available from: http://www.intechopen.com/books/principal-component-analysis/robust-principal-component-analysis-forbackground

[19] Fan Z, Yong X, Zuo W, et al. Modified principal component analysis: An integration of multiple similarity subspace models. IEEE Transactions on Neural Networks and Learning Systems. 2014;**26**(8)

[20] Jolliffe IT, Cadima J. Principal component analysis: A review and recent developments. Philosophical Transactions of Royal Society A. 2016; **374**:20150202

[21] Han Y, Song G, Liu F, Geng Z, Ma B, Xu W. Fault monitoring using novel adaptive kernel principal component analysis integrating grey relational analysis. Process Safety and Environmental Protection, Volume. 2022;**157**:397-410

[22] Gong M, Miller C, Scott M, et al. State-space functional principal component analysis to identify spatiotemporal patterns in remote sensing lake water quality. Stoch Environment Res Risk Assessment. 2021; **35**:2521-2536. DOI: 10.1007/s00477-021-02017-w

[23] Mustaqeem M, Saqib M. Principal component based support vector machine (PC-SVM): A hybrid technique for software defect detection. Cluster Computing. 2021;**24**:2581-2595

[24] Rasheed J, Hameed AA, Djeddi C, et al. A machine learning-based framework for the diagnosis of COVID-19 from chest X-ray images, Interdisciplinary Sciences: Computational. Life Sciences. 2021;**13**:103-117

[25] Lazić M, Perišić A, Perišić B. Residential Buildings Complex Boundaries Generation Based on Spatial Grid System. Appl. Sci. 2022;**12**:165. DOI: 10.3390/app12010165

[26] Available from: https://www.zachman.com/resources/ea-articles-reference/327-the-framework-for-enterprise-architecture-background-description-and-utility-by-john-a-zachman [Accessed January 10, 2022]

[27] Svyatoslav Kotusev. A comparison of top four enterprise architecture frameworks. Available from: https://www.bcs.org/articles-opinion-and-research/a-comparison-of-the-top-four-enterprise-architecture-frameworks/#16 [Accessed January 10, 2022]

[28] The Open Group. The Open Group Architecture Framework, (TOGAF). Available from: https://www.opengroup.org/togaf, https://www.opengroup.org/architecture-forum [Accessed January 1, 2022]

[29] FEA. Available from: https://www.feacinstitute.org/ [Accessed January 1, 2022]

[30] Carnegie Mellon University, Software Engineering Institute, Capability Maturity Model Integrated. Available from: https://cmmiinstitute.com/ [Accessed January 1, 2022]

[31] Cook SC, Pratt JM. Advances in systems of systems engineering foundations and methodologies. Australian Journal of Multi-Disciplinary Engineering. 2021;**17**(1):9-22. DOI: 10.1080/14488388.2020.1809845

[32] Yearworth, M., Terry, AJ, Godfrey, PS., & Edwards, G. (2010). Systems thinking research: Principles, and methodologies to grapple with complex real-world problems. In INCOSE UK Annual Systems Engineering Conference (ASEC 2010), Chipping Norton, Oxfordshire

[33] Potts MW, Sartor P, Johnson A, Bullock S. A network perspective on assessing system architectures: Foundations and challenges. Systems Engineering. 2019;**22**:485-501

[34] Salado A, Kannan H. Elemental patterns of verification strategies. Systems Engineering. 2019;**22**:370-388

[35] Knöös Franzén L, Staack I, Krus P, Jouannet C, Amadori KA. Breakdown of system of systems needs using architecture frameworks. Ontologies, and Description Logic Reasoning, Aerospace. 2021;**8**:118. DOI: 10.3390/aerospace8040118

[36] Perisic A, Lazic M, Perisic B. The extensible orchestration framework approach to collaborative design in architectural, urban and construction engineering. Automation in Construction. 2016;**71**:210-225. DOI: 10.1016/j.autcon.2016.08.005

Chapter 2

Identification of Multilinear Systems: A Brief Overview

Laura-Maria Dogariu, Constantin Paleologu, Jacob Benesty and Silviu Ciochină

Abstract

Nonlinear systems have been studied for a long time and have applications in numerous research fields. However, there is currently no global solution for nonlinear system identification, and different used approaches depend on the type of nonlinearity. An interesting class of nonlinear systems, with a wide range of popular applications, is represented by multilinear (or multidimensional) systems. These systems exhibit a particular property that may be exploited, namely that they can be regarded as linearly separable systems and can be modeled accordingly, using tensors. Examples of well-known applications of multilinear forms are multiple-input/single-output (MISO) systems and acoustic echo cancellers, used in multi-party voice communications, such as videoconferencing. Many important fields (e.g., big data, machine learning, and source separation) can benefit from the methods employed in multidimensional system identification. In this context, this chapter aims to briefly present the recent approaches in the identification of multilinear systems. Methods relying on tensor decomposition and modeling are used to address the large parameter space of such systems.

Keywords: nonlinear systems, tensor decomposition, multilinear forms, Wiener filter, adaptive filters, system identification

1. Introduction

System identification is an important topic nowadays since it can be used in solving numerous problems [1]. The aim of system identification is to estimate an unknown model using the available and observed data, namely the input and output of the system. In this context, the well-known Wiener filter is a popular solution, along with the adaptive filters which can be derived starting from this approach.

In multilinear system identification, dealing with a large parameter space represents an important challenge [2, 3]. The huge length of the filter (hundreds or thousands of coefficients) is also a serious problem [4, 5]. The methods used for addressing these issues usually rely on tensor decomposition and modeling [2, 6–17], meaning that a high-dimension problem is rewritten as a combination of lower-dimension structures, using the Kronecker product decomposition [18].

In the context of multilinear forms identification, a few approaches were proposed recently, addressing the cases when the large system is decomposed into two or three smaller components (i.e., bilinear and trilinear forms, respectively) [10, 17–23]. The aforementioned solutions outperform their conventional counterparts, offering at the same time a lower computational complexity.

Motivated by the appealing performance of these previous developments, we extended the tensor decomposition technique to higher-order systems, and in this framework, this chapter presents a part of the work and results obtained recently by the authors in the context of multilinear system identification. An iterative Wiener filter and a family of LMS-based algorithms tailored for multilinear forms are presented. For more details on the results summarized here, the works [24–26] can be consulted.

Related to the work presented here, several other tensor-based solutions relying on the recursive least-squares (RLS) algorithm were also developed recently [27, 28]. Possible applications of such system identification frameworks can be encountered in topics such as big data [29], machine learning [14], but they may be also useful in nonlinear acoustic echo cancelation [30, 31], source separation [13, 32, 33], channel equalization [12, 34], array beamforming [16, 35], blind identification [36], object recognition [37, 38], and cardiac applications [39].

The rest of this chapter is organized in the following way. In Section 2, we introduce the system model for the multiple-input/single-output (MISO) system identification problem. In this context, Section 3 presents an iterative Wiener filter tailored for the identification of multilinear systems. Next, in Section 4, an LMS-based algorithm is presented, together with its normalized version, and then, in Section 5, the performance of these algorithms is proved through simulations. Finally, conclusions are drawn in Section 6.

2. System model in the multilinear framework

Let us consider a MISO system, whose output signal at the time index t can be written as

$$y(t) = \sum_{l_1=1}^{L_1}\sum_{l_2=1}^{L_2}\cdots\sum_{l_N=1}^{L_N} x_{l_1 l_2 l_3 \dots l_N}(t) h_{1,l_1} h_{2,l_2} \cdots h_{N,l_N}, \tag{1}$$

where the individual channels are modeled by the vectors:

$$\mathbf{h}_i = \begin{bmatrix} h_{i,1} & h_{i,2} & \cdots & h_{i,L_i} \end{bmatrix}^T, \ i = 1, 2, \dots, N, \tag{2}$$

the superscript ^{T} denotes the transpose operator, and the input signals may be expressed in a tensorial form as $\boldsymbol{\mathcal{X}}(t) \in \mathbb{R}^{L_1 \times L_2 \times L_3 \times \cdots \times L_N}$, with the elements $(\boldsymbol{\mathcal{X}})_{l_1 l_2 \dots l_N}(t) = x_{l_1 l_2 \dots l_N}(t)$. Consequently, the output signal becomes

$$y(t) = \boldsymbol{\mathcal{X}}(t) \times_1 \mathbf{h}_1^T \times_2 \mathbf{h}_2^T \times_3 \cdots \times_N \mathbf{h}_N^T, \tag{3}$$

where $\times_i$ (for $i = 1, 2, \dots, N$) denotes the mode-i product [7]. It can be said that $y(t)$ is a multilinear form because it is a linear function of each of the vectors $\mathbf{h}_i, i = 1, 2, \dots, N$, when the other $N - 1$ vectors are fixed. In this context, $y(t)$ may be regarded as an extension of the bilinear form [19]. Next, let us define

$$\boldsymbol{\mathcal{H}} = \mathbf{h}_1 \circ \mathbf{h}_2 \circ \cdots \circ \mathbf{h}_N, \tag{4}$$

where $\circ$ is the vector outer product, i.e., $\mathbf{h}_1 \circ \mathbf{h}_2 = \mathbf{h}_1\mathbf{h}_2^T$, $(\mathbf{h}_1 \circ \mathbf{h}_2)_{i,j} = h_{1,i}h_{2,j}$, $\mathrm{vec}(\mathbf{h}_1 \circ \mathbf{h}_2) = \mathbf{h}_2 \otimes \mathbf{h}_1$, and

$$(\mathcal{H})_{l_1,l_2,\dots,l_N} = h_{1,l_1}h_{2,l_2}\cdots h_{N,l_N}, \tag{5}$$

$$\mathrm{vec}(\mathcal{H}) = \mathbf{h}_N \otimes \mathbf{h}_{N-1} \otimes \cdots \otimes \mathbf{h}_1, \tag{6}$$

where $\otimes$ denotes the Kronecker product and $\mathrm{vec}(\cdot)$ is the vectorization operation:

$$\mathrm{vec}(\mathcal{H}) = \begin{bmatrix} \mathrm{vec}(\mathbf{H}_{::\dots:1}) \\ \vdots \\ \mathrm{vec}(\mathbf{H}_{::\dots:L_N}) \end{bmatrix}, \tag{7}$$

$$\mathrm{vec}(\mathbf{H}_{::\dots:l_i}) = \begin{bmatrix} \mathrm{vec}(\mathbf{H}_{::\dots:1,i}) \\ \vdots \\ \mathrm{vec}\left(\mathbf{H}_{::\dots:L_{N-1},i}\right) \end{bmatrix}, \tag{8}$$

and so on, where $\mathbf{H}_{::\dots:l_i} \in \mathbb{R}^{L_1 \times L_2 \times L_3 \times \cdots \times L_{N-1}}$ represent the frontal slices of the tensor $\mathcal{H}$. Therefore, the output signal can be expressed as

$$y(t) = \mathrm{vec}^T(\mathcal{H})\mathrm{vec}[\mathcal{X}(t)], \tag{9}$$

where

$$\mathrm{vec}[\mathcal{X}(t)] = \begin{bmatrix} \mathrm{vec}[\mathbf{X}_{::\dots:1}(t)] \\ \vdots \\ \mathrm{vec}[\mathbf{X}_{::\dots:L_N}(t)] \end{bmatrix} = \mathbf{x}(t), \tag{10}$$

with $\mathbf{X}_{::\dots:l_i}(t) \in \mathbb{R}^{L_1 \times L_2 \times L_3 \times \cdots \times L_{N-1}}$ being the frontal slices of the tensor $\mathcal{X}(t)$. Let us denote the global impulse response of length $L_1L_2\cdots L_N$ as

$$\mathbf{g} = \mathrm{vec}(\mathcal{H}) = \mathbf{h}_N \otimes \mathbf{h}_{N-1} \otimes \cdots \otimes \mathbf{h}_1. \tag{11}$$

Here, an observation can be made: the solution of the decomposition in Eq. (11) is not unique [17, 24]. Despite this, no scaling ambiguity occurs in the identification of the global impulse response, $\mathbf{g}$.

Using Eqs. (9)–(11), we may rewrite $y(t)$ as

$$y(t) = \mathbf{g}^T\mathbf{x}(t). \tag{12}$$

We aim to identify the global impulse response, $\mathbf{g}$. We can define the reference (or desired) signal as

$$d(t) = \mathbf{g}^T\mathbf{x}(t) + w(t), \tag{13}$$

where $w(t)$ denotes the additive noise, which is uncorrelated with the input signals, and whose variance is

$$\sigma_d^2 = \mathbf{g}^T E\left[\mathbf{x}(t)\mathbf{x}^T(t)\right]\mathbf{g} + \sigma_w^2 = \mathbf{g}^T\mathbf{R}\mathbf{g} + \sigma_w^2, \tag{14}$$

with $E[\cdot]$ denoting mathematical expectation, $\mathbf{R} = E[\mathbf{x}(t)\mathbf{x}^T(t)]$, and $\sigma_w^2 = E[w^2(t)]$. Next, the error signal can be defined as

$$e(t) = d(t) - \hat{\mathbf{g}}^T\mathbf{x}(t), \tag{15}$$

where $\hat{\mathbf{g}}$ denotes an estimate of the global impulse response.

The optimization criterion is the minimization of the mean-squared error (MSE), which can be defined using Eq. (15):

$$J(\hat{\mathbf{g}}) = E[e^2(t)] = \sigma_d^2 - 2\hat{\mathbf{g}}^T\mathbf{p} + \hat{\mathbf{g}}^T\mathbf{R}\hat{\mathbf{g}}, \tag{16}$$

where $\mathbf{p} = E[d(t)\mathbf{x}(t)]$ denotes the cross-correlation vector between $d(t)$ and $\mathbf{x}(t)$. The solution to this minimization problem is given by the popular Wiener filter [40]:

$$\hat{\mathbf{g}}_{\mathrm{W}} = \mathbf{R}^{-1}\mathbf{p}. \tag{17}$$

Relation (17) provides the global impulse response. In order to obtain the N coefficient vectors $\mathbf{h}_i,\ i = 1, 2, \dots, N$, the nonlinear equation set containing $L_1L_2\cdots L_N$ equations with $L_1 + L_2 + \cdots + L_N$ scalar variables needs to be solved:

$$\hat{\mathbf{g}}_{\mathrm{W}} = \hat{\mathbf{h}}_{\mathrm{W},N} \otimes \hat{\mathbf{h}}_{\mathrm{W},N-1} \otimes \cdots \otimes \hat{\mathbf{h}}_{\mathrm{W},1}. \tag{18}$$

3. Multilinear iterative Wiener filter

It can be easily checked that

$$\begin{aligned}
\mathbf{g} &= \mathbf{h}_N \otimes \mathbf{h}_{N-1} \otimes \cdots \otimes \mathbf{h}_1 \\
&= (\mathbf{h}_N \otimes \mathbf{h}_{N-1} \otimes \cdots \otimes \mathbf{I}_{L_1})\mathbf{h}_1 \\
&= (\mathbf{h}_N \otimes \mathbf{h}_{N-1} \otimes \cdots \otimes \mathbf{h}_3 \otimes \mathbf{I}_{L_2} \otimes \mathbf{h}_1)\mathbf{h}_2 \\
&\ \vdots \\
&= (\mathbf{h}_N \otimes \mathbf{h}_{N-1} \otimes \cdots \otimes \mathbf{I}_{L_i} \otimes \mathbf{h}_{L_i-1} \otimes \cdots \otimes \mathbf{h}_1)\mathbf{h}_i \\
&\ \vdots \\
&= (\mathbf{I}_{L_N} \otimes \mathbf{h}_{N-1} \otimes \cdots \otimes \mathbf{h}_1)\mathbf{h}_N,
\end{aligned} \tag{19}$$

where $\mathbf{I}_{L_i}$ denotes the identity matrix of size $L_i \times L_i$.

Hence, the cost function given by Eq. (16) may be expressed in N equivalent forms:

$$\begin{aligned}
J\left(\hat{\mathbf{h}}_1, \hat{\mathbf{h}}_2, \dots, \hat{\mathbf{h}}_N\right) &= \sigma_d^2 - 2\hat{\mathbf{g}}^T\mathbf{p} + \hat{\mathbf{g}}^T\mathbf{R}\hat{\mathbf{g}} \\
&= \sigma_d^2 - 2\hat{\mathbf{h}}_i^T\left(\hat{\mathbf{h}}_N \otimes \hat{\mathbf{h}}_{N-1} \otimes \cdots \otimes \mathbf{I}_{L_i} \otimes \hat{\mathbf{h}}_{L_i-1} \otimes \cdots \otimes \hat{\mathbf{h}}_1\right)^T\mathbf{p} \\
&\quad + \hat{\mathbf{h}}_i^T\left(\hat{\mathbf{h}}_N \otimes \hat{\mathbf{h}}_{N-1} \otimes \cdots \otimes \mathbf{I}_{L_i} \otimes \hat{\mathbf{h}}_{L_i-1} \otimes \cdots \otimes \hat{\mathbf{h}}_1\right)^T \\
&\quad \times \mathbf{R}\left(\hat{\mathbf{h}}_N \otimes \hat{\mathbf{h}}_{N-1} \otimes \cdots \otimes \mathbf{I}_{L_i} \otimes \hat{\mathbf{h}}_{L_i-1} \otimes \cdots \otimes \hat{\mathbf{h}}_1\right)\hat{\mathbf{h}}_i \\
&= \sigma_d^2 - 2\hat{\mathbf{h}}_i^T\mathbf{p}_i + \hat{\mathbf{h}}_i^T\mathbf{R}_i\hat{\mathbf{h}}_i, \quad i = 1, 2, \dots, N,
\end{aligned} \tag{20}$$

where

$$\mathbf{p}_i = \left(\hat{\mathbf{h}}_N \otimes \hat{\mathbf{h}}_{N-1} \otimes \cdots \otimes \mathbf{I}_{L_i} \otimes \hat{\mathbf{h}}_{L_i-1} \otimes \cdots \otimes \hat{\mathbf{h}}_1\right)^T \mathbf{p}, \tag{21}$$

$$\begin{aligned}\mathbf{R}_i = &\left(\hat{\mathbf{h}}_N \otimes \hat{\mathbf{h}}_{N-1} \otimes \cdots \otimes \mathbf{I}_{L_i} \otimes \hat{\mathbf{h}}_{L_i-1} \otimes \cdots \otimes \hat{\mathbf{h}}_1\right)^T \mathbf{R} \\ &\times \left(\hat{\mathbf{h}}_N \otimes \hat{\mathbf{h}}_{N-1} \otimes \cdots \otimes \mathbf{I}_{L_i} \otimes \hat{\mathbf{h}}_{L_i-1} \otimes \cdots \otimes \hat{\mathbf{h}}_1\right).\end{aligned} \tag{22}$$

If all coefficients except $\hat{\mathbf{h}}_i$ are kept fixed, we may define

$$\begin{aligned}J_{\hat{\mathbf{h}}_1, \hat{\mathbf{h}}_2, \dots, \hat{\mathbf{h}}_{i-1}, \hat{\mathbf{h}}_{i+1}, \dots, \hat{\mathbf{h}}_N}\left(\hat{\mathbf{h}}_i\right) = \sigma_d^2 - 2\hat{\mathbf{h}}_i^T \mathbf{p}_i + \hat{\mathbf{h}}_i^T \mathbf{R}_i \hat{\mathbf{h}}_i, \\ i = 1, 2, \dots, N.\end{aligned} \tag{23}$$

The minimization of this convex cost function with respect to $\hat{\mathbf{h}}_i$ yields

$$\hat{\mathbf{h}}_i = \mathbf{R}_i^{-1} \mathbf{p}_i, \quad i = 1, 2, \dots, N. \tag{24}$$

Using this result, an iterative approach can be derived. A set of initial values $\hat{\mathbf{h}}_i^{(0)}, \; i = 1, 2, \dots, N$ are chosen for starting the algorithm, and then we can compute

$$\mathbf{p}_1^{(0)} = \left(\hat{\mathbf{h}}_N^{(0)} \otimes \hat{\mathbf{h}}_{N-1}^{(0)} \otimes \cdots \otimes \hat{\mathbf{h}}_2^{(0)} \otimes \hat{\mathbf{I}}_{L_1}\right)^T \mathbf{p}, \tag{25}$$

$$\mathbf{R}_1^{(0)} = \left(\hat{\mathbf{h}}_N^{(0)} \otimes \hat{\mathbf{h}}_{N-1}^{(0)} \otimes \cdots \otimes \hat{\mathbf{h}}_2^{(0)} \otimes \hat{\mathbf{I}}_{L_1}\right)^T \mathbf{R} \left(\hat{\mathbf{h}}_N^{(0)} \otimes \hat{\mathbf{h}}_{N-1}^{(0)} \otimes \cdots \otimes \hat{\mathbf{h}}_2^{(0)} \otimes \hat{\mathbf{I}}_{L_1}\right), \tag{26}$$

$$J_{\hat{\mathbf{h}}_2, \hat{\mathbf{h}}_3, \dots, \hat{\mathbf{h}}_N}\left(\hat{\mathbf{h}}_1^{(1)}\right) = \sigma_d^2 - 2\left(\hat{\mathbf{h}}_1^{(1)}\right)^T \mathbf{p}_1^{(0)} + \left(\hat{\mathbf{h}}_1^{(1)}\right)^T \mathbf{R}_1^{(0)} \left(\hat{\mathbf{h}}_1^{(1)}\right). \tag{27}$$

The minimization of the cost function yields

$$\hat{\mathbf{h}}_1^{(1)} = \left(\mathbf{R}_1^{(0)}\right)^{-1} \mathbf{p}_1^{(0)}. \tag{28}$$

Using $\hat{\mathbf{h}}_1^{(1)}$ and $\hat{\mathbf{h}}_i^{(0)}$, $i = 3, \dots, N$, we can now compute $\hat{\mathbf{h}}_2^{(1)}$. Then, the cost function becomes

$$J_{\hat{\mathbf{h}}_1, \hat{\mathbf{h}}_3, \dots, \hat{\mathbf{h}}_N}\left(\hat{\mathbf{h}}_2^{(1)}\right) = \sigma_d^2 - 2\left(\hat{\mathbf{h}}_2^{(1)}\right)^T \mathbf{p}_2^{(1)} + \left(\hat{\mathbf{h}}_2^{(1)}\right)^T \mathbf{R}_2^{(1)} \left(\hat{\mathbf{h}}_2^{(1)}\right), \tag{29}$$

where

$$\mathbf{p}_2^{(1)} = \left(\hat{\mathbf{h}}_N^{(0)} \otimes \hat{\mathbf{h}}_{N-1}^{(0)} \otimes \cdots \otimes \hat{\mathbf{h}}_3^{(0)} \otimes \hat{\mathbf{I}}_{L_2} \otimes \hat{\mathbf{h}}_1^{(1)}\right)^T \mathbf{p}, \tag{30}$$

$$\mathbf{R}_2^{(1)} = \left(\hat{\mathbf{h}}_N^{(0)} \otimes \hat{\mathbf{h}}_{N-1}^{(0)} \otimes \cdots \otimes \hat{\mathbf{h}}_3^{(0)} \otimes \hat{\mathbf{I}}_{L_2} \hat{\mathbf{h}}_1^{(1)}\right)^T \mathbf{R} \left(\hat{\mathbf{h}}_N^{(0)} \otimes \hat{\mathbf{h}}_{N-1}^{(0)} \otimes \cdots \otimes \hat{\mathbf{h}}_3^{(0)} \otimes \hat{\mathbf{I}}_{L_2} \hat{\mathbf{h}}_1^{(1)}\right). \tag{31}$$

The minimization of the cost function yields

$$\hat{\mathbf{h}}_2^{(1)} = \left(\mathbf{R}_2^{(1)}\right)^{-1} \mathbf{p}_2^{(1)}. \tag{32}$$

All the other estimates $\hat{\mathbf{h}}_i^{(1)}$, $i = 3, 4, \dots, N$ can be computed in a similar manner. By further iterating up to iteration n, the estimates of the N vectors are obtained. This minimization technique is called "block coordinate descent" [41].

4. LMS and NLMS algorithms for multilinear forms

The limitations of the Wiener filter (e.g., matrix inversion, statistics estimation) can restrict the applicability of the previously presented approach in real-world situations (for example, in nonstationary conditions, or when real-time processing is needed). Therefore, a better approach may be represented by adaptive filters. In this context, the well-known least-mean-square (LMS) algorithm is among the most popular solutions, due to its simplicity. In the following, a family of LMS-based algorithms for multilinear forms identification is presented.

By using the estimated impulse responses $\hat{\mathbf{h}}_i(t)$, $i = 1, 2, \dots, N$, the corresponding a priori error signals can be defined

$$e_{\hat{\mathbf{h}}_2\hat{\mathbf{h}}_3\dots\hat{\mathbf{h}}_N}(t) = d(t) - \hat{\mathbf{h}}_1^T(t-1)\mathbf{x}_{\hat{\mathbf{h}}_2\hat{\mathbf{h}}_3\dots\hat{\mathbf{h}}_N}(t), \tag{33}$$

$$e_{\hat{\mathbf{h}}_1\hat{\mathbf{h}}_3\dots\hat{\mathbf{h}}_N}(t) = d(t) - \hat{\mathbf{h}}_2^T(t-1)\mathbf{x}_{\hat{\mathbf{h}}_1\hat{\mathbf{h}}_3\dots\hat{\mathbf{h}}_N}(t), \tag{34}$$

$$\vdots$$

$$e_{\hat{\mathbf{h}}_1\hat{\mathbf{h}}_2\dots\hat{\mathbf{h}}_{N-1}}(t) = d(t) - \hat{\mathbf{h}}_N^T(t-1)\mathbf{x}_{\hat{\mathbf{h}}_1\hat{\mathbf{h}}_2\dots\hat{\mathbf{h}}_{N-1}}(t), \tag{35}$$

where

$$\mathbf{x}_{\hat{\mathbf{h}}_2\hat{\mathbf{h}}_3\dots\hat{\mathbf{h}}_N}(t) = [\hat{\mathbf{h}}_N(t-1)\otimes\hat{\mathbf{h}}_{N-1}(t-1)\otimes\cdots\otimes\hat{\mathbf{h}}_2(t-1)\otimes\mathbf{I}_{L_1}]\mathbf{x}(t), \tag{36}$$

$$\mathbf{x}_{\hat{\mathbf{h}}_1\hat{\mathbf{h}}_3\dots\hat{\mathbf{h}}_N}(t) = [\hat{\mathbf{h}}_N(t-1)\otimes\hat{\mathbf{h}}_{N-1}(t-1)\otimes\cdots\otimes\hat{\mathbf{h}}_3(t-1)\otimes\mathbf{I}_{L_2}\otimes\hat{\mathbf{h}}_1(t-1)]\mathbf{x}(t), \tag{37}$$

$$\vdots$$

$$\mathbf{x}_{\hat{\mathbf{h}}_1\hat{\mathbf{h}}_2\dots\hat{\mathbf{h}}_{N-1}}(t) = [\mathbf{I}_{L_N}\otimes\hat{\mathbf{h}}_{N-1}(t-1)\otimes\cdots\otimes\hat{\mathbf{h}}_2(t-1)\hat{\mathbf{h}}_1(t-1)]\mathbf{x}(t). \tag{38}$$

We can easily check that $e_{\hat{\mathbf{h}}_2\hat{\mathbf{h}}_3\dots\hat{\mathbf{h}}_N}(t) = e_{\hat{\mathbf{h}}_1\hat{\mathbf{h}}_3\dots\hat{\mathbf{h}}_N}(t) = \dots = e_{\hat{\mathbf{h}}_1\hat{\mathbf{h}}_2\dots\hat{\mathbf{h}}_{N-1}}(t)$. Hence, the LMS updates of the individual filters will be

$$\begin{aligned}\hat{\mathbf{h}}_1(t) &= \hat{\mathbf{h}}_1(t-1) - \frac{\mu_{\hat{\mathbf{h}}_1}}{2}\times\frac{\partial e^2_{\hat{\mathbf{h}}_2\hat{\mathbf{h}}_3\dots\hat{\mathbf{h}}_N}(t)}{\partial\hat{\mathbf{h}}_1(t-1)} \\ &= \hat{\mathbf{h}}_1(t-1) + \mu_{\hat{\mathbf{h}}_1}\mathbf{x}_{\hat{\mathbf{h}}_2\hat{\mathbf{h}}_3\dots\hat{\mathbf{h}}_N}(t)e_{\hat{\mathbf{h}}_2\hat{\mathbf{h}}_3\dots\hat{\mathbf{h}}_N}(t),\end{aligned} \tag{39}$$

$$\begin{aligned}\hat{\mathbf{h}}_2(t) &= \hat{\mathbf{h}}_2(t-1) - \frac{\mu_{\hat{\mathbf{h}}_2}}{2}\times\frac{\partial e^2_{\hat{\mathbf{h}}_1\hat{\mathbf{h}}_3\dots\hat{\mathbf{h}}_N}(t)}{\partial\hat{\mathbf{h}}_2(t-1)} \\ &= \hat{\mathbf{h}}_2(t-1) + \mu_{\hat{\mathbf{h}}_2}\mathbf{x}_{\hat{\mathbf{h}}_1\hat{\mathbf{h}}_3\dots\hat{\mathbf{h}}_N}(t)e_{\hat{\mathbf{h}}_1\hat{\mathbf{h}}_3\dots\hat{\mathbf{h}}_N}(t),\end{aligned} \tag{40}$$

$$\vdots$$

$$\begin{aligned}\hat{\mathbf{h}}_N(t) &= \hat{\mathbf{h}}_N(t-1) - \frac{\mu_{\hat{\mathbf{h}}_N}}{2} \times \frac{\partial e^2_{\hat{\mathbf{h}}_1\hat{\mathbf{h}}_2\ldots\hat{\mathbf{h}}_{N-1}}(t)}{\partial \hat{\mathbf{h}}_N(t-1)} \\ &= \hat{\mathbf{h}}_N(t-1) + \mu_{\hat{\mathbf{h}}_N}\mathbf{x}_{\hat{\mathbf{h}}_1\hat{\mathbf{h}}_2\ldots\hat{\mathbf{h}}_{N-1}}(t)e_{\hat{\mathbf{h}}_1\hat{\mathbf{h}}_2\ldots\hat{\mathbf{h}}_{N-1}}(t),\end{aligned} \tag{41}$$

where $\mu_{\hat{\mathbf{h}}_i} > 0,\ i = 1, 2, \ldots, N$ represent the step-size parameters. Relations (39–41) describe the LMS algorithm for multilinear forms (LMS-MF). The initialization of the estimated impulse responses could be

$$\hat{\mathbf{h}}_1(0) = [1 \quad 0 \quad \cdots \quad 0]^T, \tag{42}$$

$$\hat{\mathbf{h}}_k(0) = \frac{1}{L_k}[1 \quad 1 \quad \cdots \quad 1]^T,\ k = 2, 3, \ldots, N. \tag{43}$$

The global filter estimate is obtained as

$$\hat{\mathbf{g}}(t) = \hat{\mathbf{h}}_N(t) \otimes \hat{\mathbf{h}}_{N-1}(t) \otimes \cdots \otimes \hat{\mathbf{h}}_1(t). \tag{44}$$

We may also identify the global impulse response using the classical LMS algorithm:

$$\hat{\mathbf{g}}(t) = \hat{\mathbf{g}}(t-1) + \mu_{\hat{\mathbf{g}}}\mathbf{x}(t)e(t), \tag{45}$$

$$e(t) = d(t) - \hat{\mathbf{g}}(t-1)\mathbf{x}(t), \tag{46}$$

where $\mu_{\hat{\mathbf{g}}}$ denotes the global step-size parameter, but this would involve a single long-length adaptive filter.

When choosing the constant values of the step-size parameters from Eqs. (39)–(41), we need to consider the compromise between convergence rate and steady-state misadjustment. In certain cases, it can be more useful to have variable step-size parameters. Hence, the update equations become

$$\hat{\mathbf{h}}_1(t) = \hat{\mathbf{h}}_1(t-1) + \mu_{\hat{\mathbf{h}}_1}(t)\mathbf{x}_{\hat{\mathbf{h}}_2\hat{\mathbf{h}}_3\ldots\hat{\mathbf{h}}_N}(t)e_{\hat{\mathbf{h}}_2\hat{\mathbf{h}}_3\ldots\hat{\mathbf{h}}_N}(t), \tag{47}$$

$$\hat{\mathbf{h}}_2(t) = \hat{\mathbf{h}}_2(t-1) + \mu_{\hat{\mathbf{h}}_2}(t)\mathbf{x}_{\hat{\mathbf{h}}_1\hat{\mathbf{h}}_3\ldots\hat{\mathbf{h}}_N}(t)e_{\hat{\mathbf{h}}_1\hat{\mathbf{h}}_3\ldots\hat{\mathbf{h}}_N}(t), \tag{48}$$

$$\vdots$$

$$\hat{\mathbf{h}}_N(t) = \hat{\mathbf{h}}_N(t-1) + \mu_{\hat{\mathbf{h}}_N}(t)\mathbf{x}_{\hat{\mathbf{h}}_1\hat{\mathbf{h}}_2\ldots\hat{\mathbf{h}}_{N-1}}(t)e_{\hat{\mathbf{h}}_1\hat{\mathbf{h}}_2\ldots\hat{\mathbf{h}}_{N-1}}(t). \tag{49}$$

Then, the a posteriori error signals can be defined as

$$\varepsilon_{\hat{\mathbf{h}}_2\hat{\mathbf{h}}_3\ldots\hat{\mathbf{h}}_N}(t) = d(t) - \hat{\mathbf{h}}_1^T(t)\mathbf{x}_{\hat{\mathbf{h}}_2\hat{\mathbf{h}}_3\ldots\hat{\mathbf{h}}_N}(t), \tag{50}$$

$$\varepsilon_{\hat{\mathbf{h}}_1\hat{\mathbf{h}}_3\ldots\hat{\mathbf{h}}_N}(t) = d(t) - \hat{\mathbf{h}}_2^T(t)\mathbf{x}_{\hat{\mathbf{h}}_1\hat{\mathbf{h}}_3\ldots\hat{\mathbf{h}}_N}(t), \tag{51}$$

$$\vdots$$

$$\varepsilon_{\hat{\mathbf{h}}_1\hat{\mathbf{h}}_2\ldots\hat{\mathbf{h}}_{N-1}}(t) = d(t) - \hat{\mathbf{h}}_N^T(t)\mathbf{x}_{\hat{\mathbf{h}}_1\hat{\mathbf{h}}_2\ldots\hat{\mathbf{h}}_{N-1}}(t). \tag{52}$$

After replacing Eq. (39) in Eq. (50), Eq. (40) in Eq. (51), and Eq. (41) in Eq. (52), and then canceling the a posteriori error signals, we get

$$e_{\hat{\mathbf{h}}_2\hat{\mathbf{h}}_3\ldots\hat{\mathbf{h}}_N}(t)\left[1-\mu_{\hat{\mathbf{h}}_1}(t)\mathbf{x}^T_{\hat{\mathbf{h}}_2\hat{\mathbf{h}}_3\ldots\hat{\mathbf{h}}_N}(t)\mathbf{x}_{\hat{\mathbf{h}}_2\hat{\mathbf{h}}_3\ldots\hat{\mathbf{h}}_N}(t)\right]=0, \tag{53}$$

$$e_{\hat{\mathbf{h}}_1\hat{\mathbf{h}}_3\ldots\hat{\mathbf{h}}_N}(t)\left[1-\mu_{\hat{\mathbf{h}}_2}(t)\mathbf{x}^T_{\hat{\mathbf{h}}_1\hat{\mathbf{h}}_3\ldots\hat{\mathbf{h}}_N}(t)\mathbf{x}_{\hat{\mathbf{h}}_1\hat{\mathbf{h}}_3\ldots\hat{\mathbf{h}}_N}(t)\right]=0, \tag{54}$$

$$\vdots$$

$$e_{\hat{\mathbf{h}}_1\hat{\mathbf{h}}_2\ldots\hat{\mathbf{h}}_{N-1}}(t)\left[1-\mu_{\hat{\mathbf{h}}_N}(t)\mathbf{x}^T_{\hat{\mathbf{h}}_1\hat{\mathbf{h}}_2\ldots\hat{\mathbf{h}}_{N-1}}(t)\mathbf{x}_{\hat{\mathbf{h}}_1\hat{\mathbf{h}}_2\ldots\hat{\mathbf{h}}_{N-1}}(t)\right]=0. \tag{55}$$

We assume that $e_{\hat{\mathbf{h}}_2\hat{\mathbf{h}}_3\ldots\hat{\mathbf{h}}_N}(t)\neq 0$, $e_{\hat{\mathbf{h}}_1\hat{\mathbf{h}}_3\ldots\hat{\mathbf{h}}_N}(t)\neq 0$, …, $e_{\hat{\mathbf{h}}_1\hat{\mathbf{h}}_2\ldots\hat{\mathbf{h}}_{N-1}}(t)\neq 0$. Hence, the step-sizes become

$$\mu_{\hat{\mathbf{h}}_1}(t)=\frac{1}{\mathbf{x}^T_{\hat{\mathbf{h}}_2\hat{\mathbf{h}}_3\ldots\hat{\mathbf{h}}_N}(t)\mathbf{x}_{\hat{\mathbf{h}}_2\hat{\mathbf{h}}_3\ldots\hat{\mathbf{h}}_N}(t)}, \tag{56}$$

$$\mu_{\hat{\mathbf{h}}_2}(t)=\frac{1}{\mathbf{x}^T_{\hat{\mathbf{h}}_1\hat{\mathbf{h}}_3\ldots\hat{\mathbf{h}}_N}(t)\mathbf{x}_{\hat{\mathbf{h}}_1\hat{\mathbf{h}}_3\ldots\hat{\mathbf{h}}_N}(t)}, \tag{57}$$

$$\vdots$$

$$\mu_{\hat{\mathbf{h}}_N}(t)=\frac{1}{\mathbf{x}^T_{\hat{\mathbf{h}}_1\hat{\mathbf{h}}_2\ldots\hat{\mathbf{h}}_{N-1}}(t)\mathbf{x}_{\hat{\mathbf{h}}_1\hat{\mathbf{h}}_2\ldots\hat{\mathbf{h}}_{N-1}}(t)}. \tag{58}$$

In the numerators of Eqs. (56)–(58), the normalized step-size parameters $0<\alpha_{\hat{\mathbf{h}}_i}<1,\ i=1,2,\ldots,N$ can be used to achieve a good compromise between convergence rate and misadjustment. Some regularization constants denoted by $\delta_{\hat{\mathbf{h}}_i}>0,\ i=1,2,\ldots,N$ are also introduced in the denominators of the step-sizes in order to ensure robust adaptation. Consequently, we obtain the update equations of the normalized LMS (NLMS) algorithm for multilinear forms (NLMS-MF):

$$\hat{\mathbf{h}}_1(t)=\hat{\mathbf{h}}_1(t-1)+\frac{\alpha_{\hat{\mathbf{h}}_1}\mathbf{x}_{\hat{\mathbf{h}}_2\hat{\mathbf{h}}_3\ldots\hat{\mathbf{h}}_N}(t)e_{\hat{\mathbf{h}}_2\hat{\mathbf{h}}_3\ldots\hat{\mathbf{h}}_N}(t)}{\delta_{\hat{\mathbf{h}}_1}+\mathbf{x}^T_{\hat{\mathbf{h}}_2\hat{\mathbf{h}}_3\ldots\hat{\mathbf{h}}_N}(t)\mathbf{x}_{\hat{\mathbf{h}}_2\hat{\mathbf{h}}_3\ldots\hat{\mathbf{h}}_N}(t)}, \tag{59}$$

$$\hat{\mathbf{h}}_2(t)=\hat{\mathbf{h}}_2(t-1)+\frac{\alpha_{\hat{\mathbf{h}}_2}\mathbf{x}_{\hat{\mathbf{h}}_1\hat{\mathbf{h}}_3\ldots\hat{\mathbf{h}}_N}(t)e_{\hat{\mathbf{h}}_1\hat{\mathbf{h}}_3\ldots\hat{\mathbf{h}}_N}(t)}{\delta_{\hat{\mathbf{h}}_2}+\mathbf{x}^T_{\hat{\mathbf{h}}_1\hat{\mathbf{h}}_3\ldots\hat{\mathbf{h}}_N}(t)\mathbf{x}_{\hat{\mathbf{h}}_1\hat{\mathbf{h}}_3\ldots\hat{\mathbf{h}}_N}(t)}, \tag{60}$$

$$\vdots$$

$$\hat{\mathbf{h}}_N(t)=\hat{\mathbf{h}}_N(t-1)+\frac{\alpha_{\hat{\mathbf{h}}_N}\mathbf{x}_{\hat{\mathbf{h}}_1\hat{\mathbf{h}}_2\ldots\hat{\mathbf{h}}_{N-1}}(t)e_{\hat{\mathbf{h}}_1\hat{\mathbf{h}}_2\ldots\hat{\mathbf{h}}_{N-1}(t)}}{\delta_{\hat{\mathbf{h}}_N}+\mathbf{x}^T_{\hat{\mathbf{h}}_1\hat{\mathbf{h}}_2\ldots\hat{\mathbf{h}}_{N-1}}(t)\mathbf{x}_{\hat{\mathbf{h}}_1\hat{\mathbf{h}}_2\ldots\hat{\mathbf{h}}_{N-1}}(t)}. \tag{61}$$

The initialization of the individual impulse responses can be done using Eqs. (42, 43). We may also identify the global impulse response using the regular NLMS algorithm:

$$\hat{\mathbf{g}}(t)=\hat{\mathbf{g}}(t-1)+\frac{\alpha_{\hat{\mathbf{g}}}\mathbf{x}(t)e(t)}{\mathbf{x}^T(t)\mathbf{x}(t)+\delta_{\hat{\mathbf{g}}}}, \tag{62}$$

where $0<\alpha_{\hat{\mathbf{g}}}\leq 1$ denotes the normalized step-size parameter, $\delta_{\hat{\mathbf{g}}}>0$ represents the regularization constant, and $e(t)$ is defined in Eq. (46). However, this would involve a single (long length) adaptive filter, with $L=L_1L_2\cdots L_N$.

5. Experimental results

The purpose of this section is to illustrate through simulations the improved performance of the proposed solutions for the identification of multilinear forms. We performed experiments involving MISO system identification. As input signals, we used white Gaussian noises and AR(1) processes, obtained by filtering white Gaussian noises through a first-order system with the transfer function $1/(1-0.99z^{-1})$. The noise is white and Gaussian, having the variance equal to $\sigma_w^2 = 0.01$. We use two different orders of the system ($N = 4$ and $N = 5$). The first individual impulse response, $\mathbf{h}_1$, of length $L_1 = 16$, is formed using the first 16 coefficients of the first network echo path from the ITU-T G168 Recommendation [42]. The second impulse response, $\mathbf{h}_2$, of length $L_2 = 8$, is randomly generated from a Gaussian distribution, whereas the other three impulse responses, $\mathbf{h}_3$, $\mathbf{h}_4$, and $\mathbf{h}_5$, of lengths $L_3 = L_4 = 4$ and $L_5 = 2$, respectively, are obtained using an exponential decay based on the rule $h_{j,l_j} = a_j^{l_j-1}$, with $j = 3, 4, 5$, where a_j takes the values 0.9, 0.5, and 0.1, respectively. Hence, the global impulse responses $\mathbf{g}$, computed using Eq. (11), have lengths 2048, when $N = 4$, and 4096, when $N = 5$. **Figure 1** illustrates the individual impulse responses $\mathbf{h}_1$, $\mathbf{h}_2$, $\mathbf{h}_3$, and $\mathbf{h}_4$, as well as the resulting global impulse response $\mathbf{g} = \mathbf{h}_4 \otimes \mathbf{h}_3 \otimes \mathbf{h}_2 \otimes \mathbf{h}_1$, in the case when $N = 4$.

The measure of performance is the normalized misalignment (in dB) for the identification of the global impulse response, computed as $20\log_{10}\left[\|\mathbf{g} - \hat{\mathbf{g}}(n)\|_2 / \|\mathbf{g}\|_2\right]$.

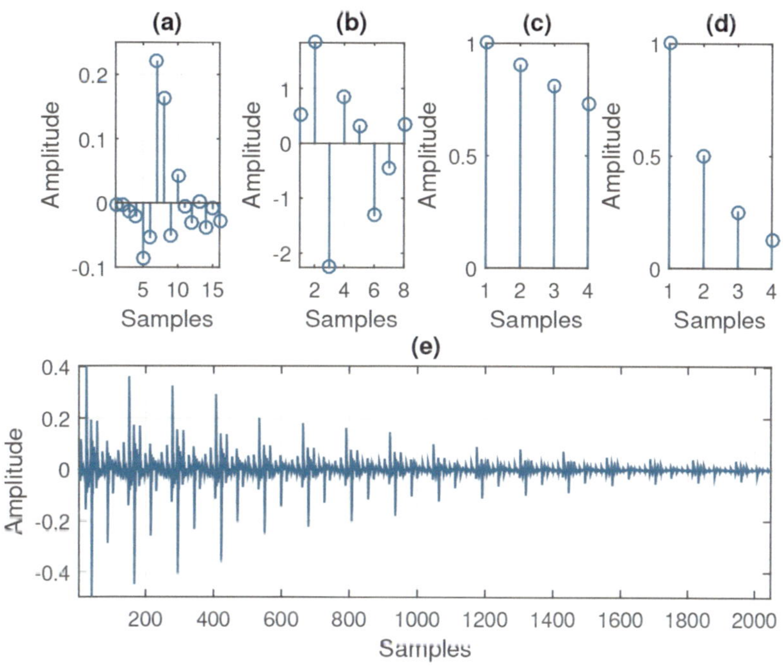

Figure 1.
*Impulse responses used in simulations of the multiple-input/single-output (MISO) system identification scenario (for $N = 4$): (**a**) $\mathbf{h}_1$ (of length $L_1 = 16$) contains the first 16 coefficients of the first impulse response from G168 recommendation [42], (**b**) $\mathbf{h}_2$ (of length $L_2 = 8$) is a randomly generated impulse response, (**c**) $\mathbf{h}_3$ (of length $L_3 = 4$) has the coefficients computed as $h_{3,l_3} = 0.9^{l_3-1}$, with $l_3 = 1, 2, \ldots, L_3$, (**d**) $\mathbf{h}_4$ (of length $L_4 = 4$) has the coefficients computed as $h_{4,l_4} = 0.5^{l_4-1}$, with $l_4 = 1, 2, \ldots, L_4$, and (**e**) $\mathbf{g}$ (of length $L = L_1L_2L_3L_4 = 2048$) is the global impulse response, which results based on Eq. (11).*

A sudden change in the sign of the coefficients of $\mathbf{h}_1$ is introduced in the middle of each experiment, with the goal of observing the tracking capabilities of the proposed algorithms.

First, we aim to show comparatively the performances of the LMS-MF and LMS algorithms. When choosing the step-size parameter values, we need to take into account the theoretical upper bound, which for the conventional LMS is $2/\left(L\sigma_x^2\right)$, where σ_x^2 is the input signal's variance [40]. In practical scenarios, this limit may not be usable, due to stability issues. The step-size parameters for the LMS-MF and LMS algorithms were chosen in our simulations in such a manner that similar misalignment values are obtained, in order to compare their convergence rate and tracking. The largest value of the step-size parameter for the conventional LMS algorithm is chosen close to its stability limit, such that the fastest convergence possible is obtained.

Figure 2 shows the case when $N = 4$ and the input signals are white Gaussian noises. The resulting global filter length is $L = 2048$. It can be seen that the LMS-MF achieves a higher convergence rate and faster tracking as compared to the conventional LMS algorithm. Of course, when the value of the step-size parameter decreases, the misalignment decreases, but at the same time, the convergence becomes slower.

Next, in **Figure 3**, $N = 4$ and the input signals are highly-correlated AR(1) processes. The performance gain of the LMS-MF with respect to the conventional LMS algorithm in terms of both convergence rate/tracking and steady-state misalignment is even higher in this scenario.

When the system order increases, the improvement in performance brought by the LMS-MF is even more apparent. This can be seen in **Figure 4**, where $N = 5$ and the input signals are AR(1) processes. The resulting global impulse response length is $L = 4096$. It is easily observed that the proposed algorithm is superior to the conventional LMS in terms of both convergence rate/tracking and misalignment.

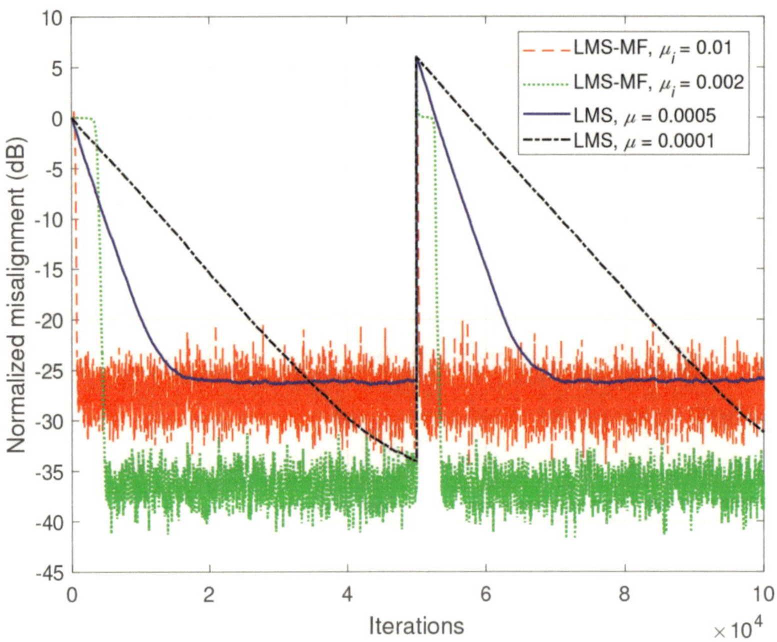

Figure 2.
Performance of the LMS-MF and LMS algorithms (using different step-size parameters), for the identification of the global impulse response **g**. *The input signals are white Gaussian noises, $N = 4$, and $L = 2048$.*

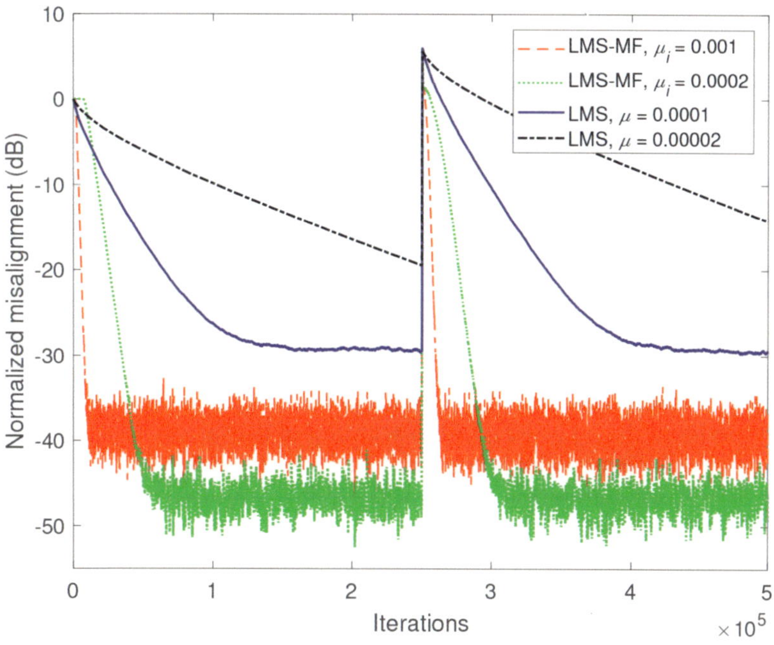

Figure 3.
Performance of the LMS-MF and LMS algorithms (using different step-size parameters), for the identification of the global impulse response **g**. *The input signals are AR(1) processes,* $N = 4$, *and* $L = 2048$.

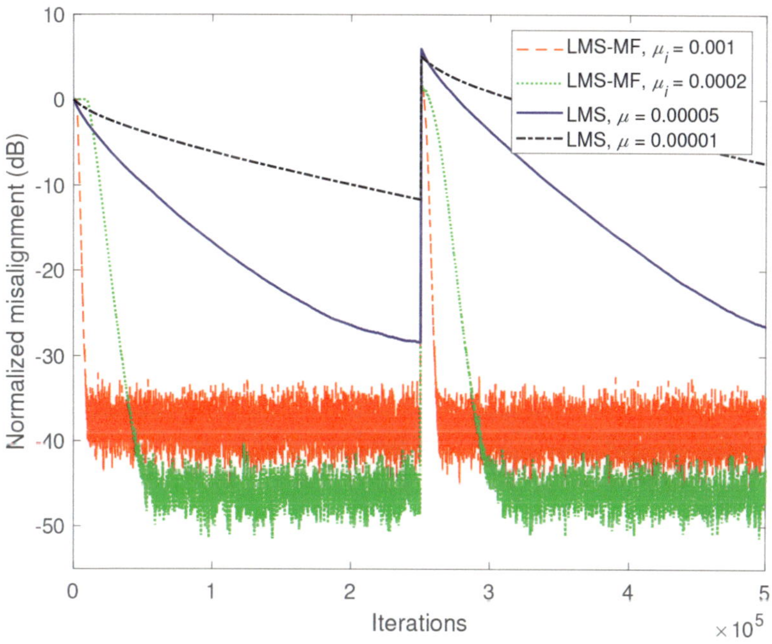

Figure 4.
Performance of the LMS-MF and LMS algorithms (using different step-size parameters), for the identification of the global impulse response **g**. *The input signals are AR(1) processes,* $N = 5$, *and* $L = 4096$.

In the following, we aim to illustrate the performance of the NLMS-MF and NLMS algorithms in the identification of the global system. Since the step-size parameter does no longer have a constant value, the normalized algorithms can work better in nonstationary environments. The fastest-convergence bound for the value of the normalized step-size parameter of the conventional NLMS algorithm is 1 [40].

Figure 5 illustrates the case when the inputs are white Gaussian noises and $N = 4$, leading to a length of the global system of $L = 2048$. Similar to the case of the LMS-MF and LMS algorithms from **Figure 2**, it can be concluded that the performance of the NLMS-MF algorithm is significantly better than the one of its conventional counterparts. This is even more apparent in the case of smaller normalized step-size values.

The improvement offered by the proposed approach is even more significant for correlated inputs. In **Figure 6**, the input signals are AR(1) processes. It is noticed that even when the NLMS-MF algorithm uses lower values for the normalized step-sizes, it can still outperform the NLMS algorithm working in the fastest convergence mode.

The same conclusion applies when the order N is increased. **Figure 7** shows the case when $N = 5$ and, hence, the length of the global impulse response is $L = 4096$. Again, the NLMS-MF algorithm achieves a significantly better convergence rate and tracking with respect to the conventional NLMS algorithm.

Next, we aim to show the influence of the normalized step-size values on the performance of the proposed algorithm. In **Figure 8**, the order of the system is $N = 4$ and the input signals are AR(1) processes. As it was expected, lower values of the normalized step-sizes improve the misalignment, but at the cost of a slower convergence and tracking.

The last experiment involving the NLMS-MF algorithm aims to show the performance in the case when the normalized step-size parameters α_i ($i = 1, 2, \dots, N$) take different values for each individual filter. In **Figure 9**, the value of the normalized

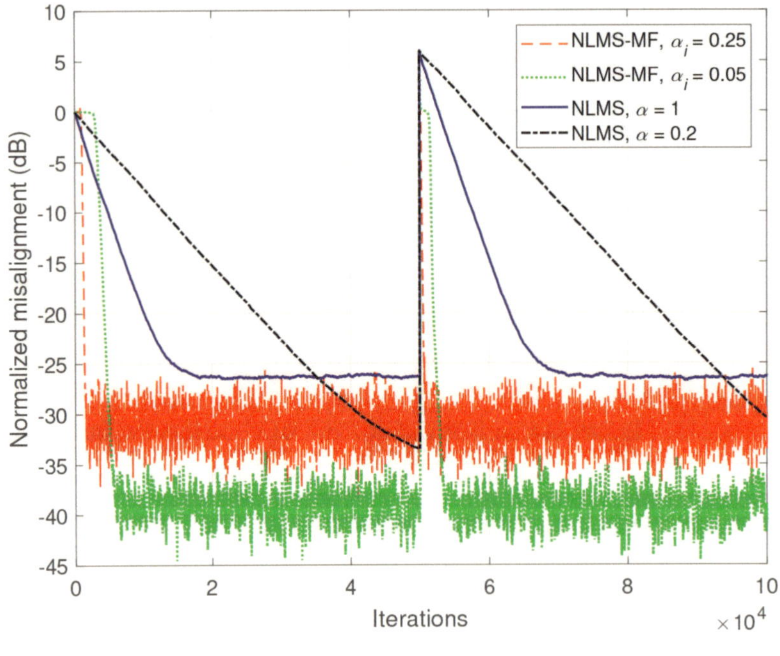

Figure 5.
Performance of the NLMS-MF and NLMS algorithms (using different normalized step-size parameters), for the identification of the global impulse response **g**. *The input signals are white Gaussian noises,* $N = 4$, *and* $L = 2048$.

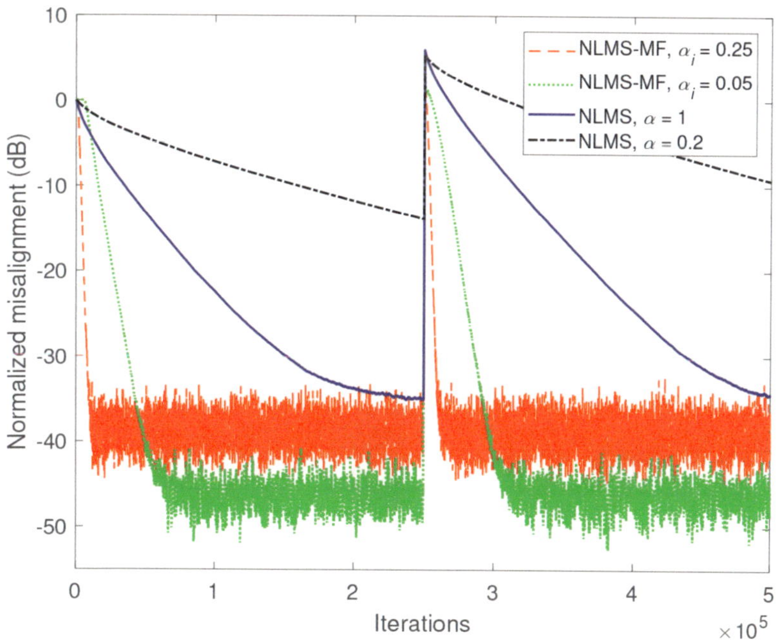

Figure 6.
Performance of the NLMS-MF and NLMS algorithms (using different normalized step-size parameters), for the identification of the global impulse response **g**. *The input signals are AR(1) processes, $N = 4$, and $L = 2048$.*

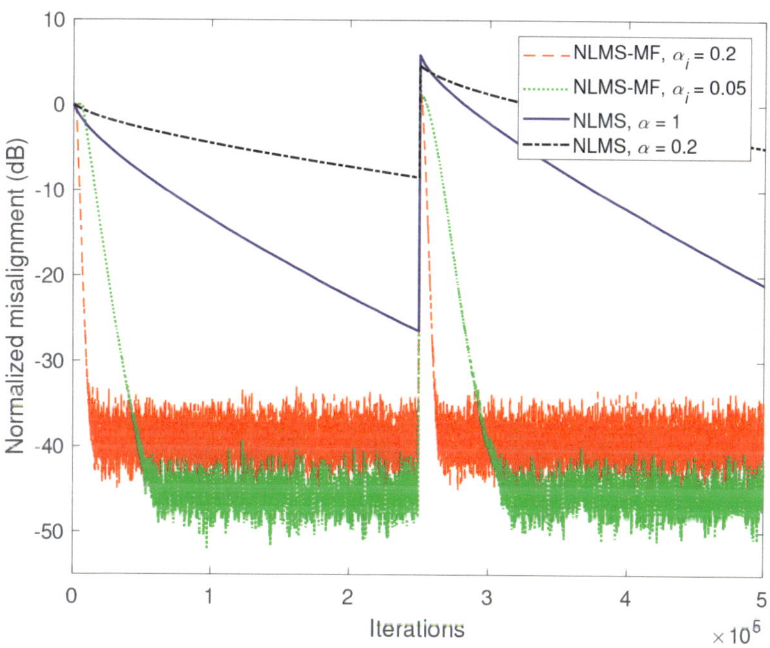

Figure 7.
Performance of the NLMS-MF and NLMS algorithms (using different normalized step-size parameters), for the identification of the global impulse response **g**. *The input signals are AR(1) processes, $N = 5$, and $L = 4096$.*

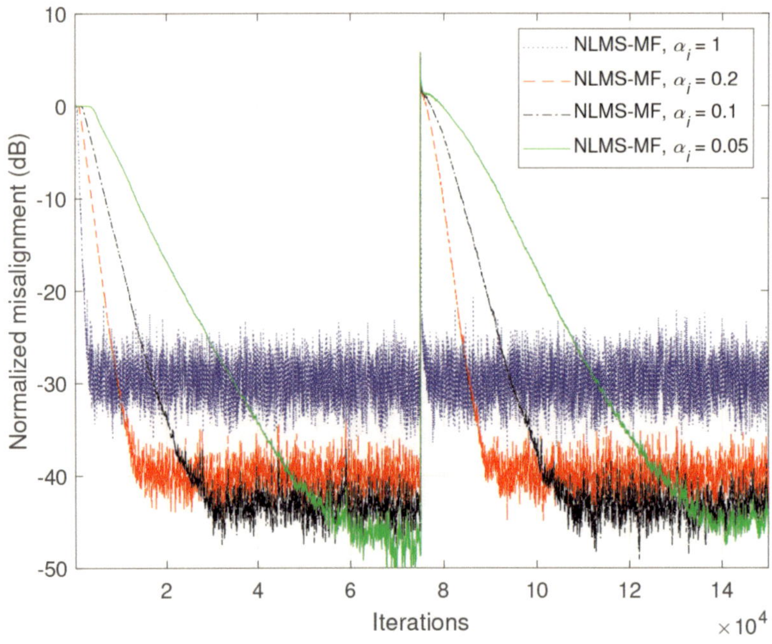

Figure 8.
Performance of the NLMS-MF algorithm using different normalized step-size parameters (with equal values of α_i, $i = 1, 2, \ldots, N$)*, for the identification of the global impulse response* $\mathbf{g}$*. The input signals are AR(1) processes,* $N = 4$*, and* $L = 2048$.

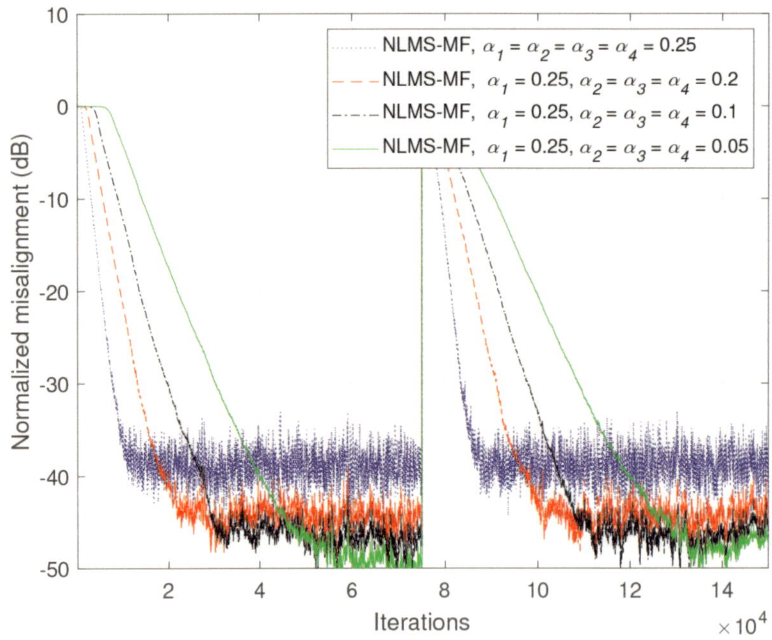

Figure 9.
Performance of the NLMS-MF algorithm using different normalized step-size parameters (with different values of α_i, $i = 1, 2, \ldots, N$)*, for the identification of the global impulse response* $\mathbf{g}$*. The input signals are AR(1) processes,* $N = 4$*, and* $L = 2048$.

step-size parameter for the first filter, of highest length, L_1, is $\alpha_1 = 0.25$, whereas the other normalized step-sizes $\alpha_j, \ j = 2, 3, 4$ are varied. The system order is $N = 4$ and the input signals are AR(1) processes. The compromise between convergence rate and misalignment can be seen again. Since the convergence is mostly influenced by the individual filter with the highest length [20], different values of the normalized step-sizes may be used in different scenarios.

Due to the important improvement in performance brought by the adaptive tensor-based LMS algorithms, observed through experiments, these algorithms may represent appealing solutions for the identification of long-length separable system impulse responses.

6. Conclusions

In this chapter, we have presented a decomposition-based approach for dealing with the identification of high-dimension MISO systems. Unlike the conventional method, which is based on the identification of the global system impulse response, our solution focuses on regarding the system as an N-order tensor and thus estimating N shorter filters. At the end, the solutions are combined ("tensorized" together) into the original high-dimension tensor. Based on the tensor decomposition technique, an iterative Wiener filter was proposed, along with a family of LMS-based algorithms suitable for the identification of such systems, also called multilinear systems. In addition to the lower computational complexity, the proposed solutions achieve superior performance as compared to their conventional counterparts from the point of view of convergence rate, tracking capability, and steady-state misadjustment. Experiments have also shown the performance improvement of the proposed adaptive algorithms for multilinear forms in long-length system identification in different scenarios, even for highly correlated input signals.

Acknowledgements

This work was supported by a grant from the Romanian Ministry of Education and Research, CNCS–UEFISCDI, Project Number PN-III-P1-1.1-PD-2019-0340, within PNCDI III.

Author details

Laura-Maria Dogariu[1*], Constantin Paleologu[1], Jacob Benesty[2] and Silviu Ciochină[1]

1 Department of Telecommunications, University Politehnica of Bucharest, Romania

2 INRS-EMT, University of Quebec, Montreal, Canada

*Address all correspondence to: ldogariu@comm.pub.ro

References

[1] Ljung L. System Identification: Theory for the User. 2nd ed. Upper Saddle River, NJ, USA: Prentice-Hall; 1999

[2] Rupp M, Schwarz S. A tensor LMS algorithm. In: Proceedings of IEEE ICASSP. Brisbane, Queensland, Australia: ICASSP; 2015. pp. 3347-3351

[3] Rupp M, Schwarz S. Gradient-based approaches to learn tensor products. In: Proc. EUSIPCO. Nice, France: EUSIPCO; 2015. pp. 2486-2490

[4] Gay SL, Benesty J. Acoustic Signal Processing for Telecommunication. Boston, MA, USA: Kluwer Academic Publisher; 2000

[5] Benesty J, GAnsler T, Morgan DR, Sondhi MM, Gay SL. Advances in Network and Acoustic Echo Cancellation. Berlin, Germany: Springer-Verlag; 2001

[6] Lathauwer LD. Signal Processing Based on Multilinear Algebra. Leuven, Belgium: Katholieke Universiteit Leuven; 1997

[7] Kolda TG, Bader BW. Tensor decompositions and applications. SIAM Review. 2009;**51**:455-500

[8] Comon P. Tensors: A brief introduction. IEEE Signal Processing Magazine. 2014;**31**:44-53

[9] Cichocki A, Mandic DP, Huyphan A, Caiafa CF, Zhou G, Zhao Q, et al. Tensor decompositions for signal processing applications. IEEE Signal Processing Magazine. 2015;**32**:145-163

[10] Ribeiro LN, de Almeida ALF, Mota JCM. Identification of Separable Systems Using Trilinear Filtering. Kos, Greece: Proceedings of IEEE CAMSAP; 2015. pp. 189-192

[11] Da Silva AP, Comon P, de Almeida ALF. A finite algorithm to compute rank-1 tensor approximations. IEEE Signal Processing Letters. 2016;**23**: 959-963

[12] Ribeiro LN, Schwarz S, Rupp M, de Almeida ALF, Mota JCM. A low-complexity equalizer for massive MIMO systems based on array separability. Kos, Greece: Proceedings of EUSIPCO; 2017. pp. 2522-2526

[13] Boussé M, Debals O, De Lathauwer L. A tensor-based method for large-scale blind source separation using segmentation. IEEE Transactions on Signal Processing. 2017;**65**(2):346-358

[14] Sidiropoulos N, De Lathauwer L, Fu X, Huang K, Papalexakis E, Faloutsos C. Tensor decomposition for signal processing and machine learning. IEEE Transactions on Signal Processing. 2017;**65**(13):3551-3582

[15] da Costa MN, Favier G, Romano JMT. Tensor modelling of MIMO communication systems with performance analysis and Kronecker receivers. Signal Processing. 2018;**145**:304-316

[16] Ribeiro LN, de Almeida ALF, Mota JCM. Separable linearly constrained minimum variance beamformers. Signal Processing. 2019; **158**:15-25

[17] Dogariu L-M, Ciochină S, Benesty J, Paleologu C. System identification based on tensor decompositions: A trilinear approach. Symmetry. 2019;**11**:556

[18] Van Loan CF. The ubiquitous Kronecker product. Journal of

Computational Applied Mathematics. 2000;**123**:85-100

[19] Benesty J, Paleologu C, Ciochinˇa S. On the identification of bilinear forms with the wiener filter. IEEE Signal Processing Letters. 2017;**24**:653-657

[20] Paleologu C, Benesty J, Ciochină S. Adaptive filtering for the identification of bilinear forms. Digital Signal Processing. 2018;**75**:153-167

[21] Elisei-Iliescu C, Stanciu C, Paleologu C, Benesty J, Anghel C, Ciochina S. Efficient recursive least-squares algorithms for the identification of bilinear forms. Digital Signal Processing. 2018;**83**:280-296

[22] Dogariu L, Paleologu C, Ciochină S, Benesty J, Piantanida P. Identification of bilinear forms with the Kalman filter. In: Proceedings IEEE International Conference on Acoustics, Speech and Signal Processing (ICASSP). Calgary, AB, Canada: IEEE; 2018. pp. 4134-4138

[23] Elisei-Iliescu C, Dogariu L-M, Paleologu C, Benesty J, Enescu AA, Ciochina S. A recursive least-squares algorithm for the identification of trilinear forms. Algorithms. 2020;**13**:135

[24] Dogariu L-M, Ciochină S, Paleologu C, Benesty J, Oprea C. An Iterative Wiener Filter for the Identification of Multilinear Forms. Milan, Italy: Proceedings of IEEE TSP; 2020. pp. 193-197

[25] Dogariu L-M, Paleologu C, Benesty J, Oprea C, Ciochină S. LMS Algorithms for Multilinear Forms. Timisoara, Romania: Proceedings of IEEE ISETC; 2020. pp. 1-4

[26] Dogariu L-M, Paleologu C, Benesty J, Oprea C, Ciochină S. Tensor-based adaptive filtering algorithms. Symmetry. 2021;**13**(3):481

[27] Fîciu ID, Stanciu C-L, Anghel C, Elisei-Iliescu C. Low-complexity Recursive Least-Squares adaptive algorithm based on tensorial forms. Applied Sciences. 2021;**11**(18):8656

[28] Fîciu ID, Stanciu C-L, Elisei-Iliescu C, Anghel C. Tensor-based Recursive Least-Squares adaptive algorithms with low-complexity and high robustness features. Electronics. 2022;**11**(2):237

[29] Vervliet N, Debals O, Sorber L, De Lathauwer L. Breaking the curse of dimensionality using decompositions of incomplete tensors: Tensor-based scientific computing in big data analysis. IEEE Signal Processing Magazine. 2014; **31**:71-79

[30] Stenger A, Kellermann W. Adaptation of a memoryless preprocessor for nonlinear acoustic echo cancelling. Signal Processing. 2000;**80**: 1747-1760

[31] Huang Y, Skoglund J, Luebs A. Practically efficient nonlinear acoustic echo cancellers using cascaded block RLS and FLMS adaptive filters. In: Proceedings of the 2017 IEEE International Conference on Acoustics, Speech and Signal Processing (ICASSP). New Orleans, LA, USA: IEEE; 2017. pp. 596-600

[32] Cichocki A, Zdunek R, Pan AH, Amari S. Nonnegative Matrix and Tensor Factorizations: Applications to Exploratory Multiway Data Analysis and Blind Source Separation. Chichester, UK: Wiley; 2009

[33] Domanov I, Lathauwer LD. From computation to comparison of tensor decompositions. SIAM Journal on Matrix Analysis and Applications. 2021;**42**(2): 449-474

[34] Gesbert D, Duhamel P. Robust blind joint data/channel estimation based on

bilinear optimization. In: Proceedings of the 8th Workshop on Statistical Signal and Array Processing. Corfu, Greece: IEEE; 1996. pp. 168-171

[35] Benesty J, Cohen I, Chen J. Array Processing–Kronecker Product Beamforming. Cham, Switzerland: Springer; 2019

[36] Ayvaz M, De Lathauwer L. Tensor-based Multivariate Polynomial Optimization with Application in Blind Identification. Dublin, Ireland: Proceedings of EUSIPCO; 2021

[37] Vasilescu MAO, Kim E. Compositional Hierarchical Tensor Factorization: Representing Hierarchical Intrinsic and Extrinsic Causal Factors. Anchorage, AK, USA: Proceedings of ACM SIGKDD; 2019

[38] Vasilescu MAO, Kim E, Zeng XS. CausalX: Causal eXplanations and Block Multilinear Factor Analysis. Milan, Italy: Proceedings of IEEE ICPR; 2021

[39] Padhy S, Goovaerts G, Boussé M, De Lathauwer L, Van Huffel S. The Power of Tensor-Based Approaches in Cardiac Applications. Singapore: Springer; 2020

[40] Haykin S. Adaptive Filter Theory. Fourth ed. Upper Saddle River, NJ: Prentice-Hall; 2002

[41] Bertsekas D. Nonlinear Programming. 2nd ed. Belmont, Massachusetts: Athena Scientific; 1999

[42] Digital Network Echo Cancellers. ITU-T Recommendations G.168. Geneva, Switzerland: WHO; 2002

Chapter 3

Evaluation of Principal Component Analysis Variants to Assess Their Suitability for Mobile Malware Detection

Padmavathi Ganapathi, Shanmugapriya Dhathathri and Roshni Arumugam

Abstract

Principal component analysis (PCA) is an unsupervised machine learning algorithm that plays a vital role in reducing the dimensions of the data in building an appropriate machine learning model. It is a statistical process that transforms the data containing correlated features into a set of uncorrelated features with the help of orthogonal transformations. Unsupervised machine learning is a concept of self-learning method that involves unlabelled data to identify hidden patterns. PCA converts the data features from a high dimensional space into a low dimensional space. PCA also acts as a feature extraction method since it transforms the '*n*' number of features into '*m*' number of principal components (PCs; $m < n$). Mobile Malware is increasing tremendously in the digital era due to the growth of android mobile users and android applications. Some of the mobile malware are viruses, Trojan horses, worms, adware, spyware, ransomware, riskware, banking malware, SMS malware, keylogger, and many more. To automate the process of detecting mobile malware without human intervention, machine learning methods are applied to discover the malware more precisely. Specifically, unsupervised machine learning helps to uncover the hidden patterns to detect anomalies in the data. In discovering hidden patterns of malware, PCA is an important dimensionality reduction technique that can be applied to transform the features into PCs containing important feature values. So, by implementing PCA, the correlated features are transformed into uncorrelated features automatically to explore the anomalies in the data effectively. This book chapter explains all the variants of the PCA, including all linear and non-linear methods of PCA and their suitability in applying to mobile malware detection. A case study on mobile malware detection with variants of PCA using machine learning techniques in CICMalDroid_2020 dataset has been experimented. Based on the experimental results, for the given dataset, normal PCA is suitable to detect the malware data points and forms an optimal cluster.

Keywords: cyber security, dimensionality reduction, machine learning, mobile malware, principal component analysis, variants of PCA

IntechOpen

1. Introduction

PCA is a statistical technique for compressing the content of large datasets into a smaller number of summary indices that can be examined and evaluated more quickly. Principal component analysis (PCA) is a multivariate statistical methodology that is frequently utilized nowadays [1]. It is a factor analysis-based statistical method that is widely used in the disciplines of pattern recognition and signal processing. It is a dimensionality reduction technique that condenses a large number of variables into a smaller set while maintaining the majority of the larger dataset. Because smaller datasets are easier to examine and visualize, machine learning algorithms can assess data more efficiently and rapidly without dealing with extra impediments.

PCA is also commonly employed in exploratory data analysis and prediction model construction. It is frequently used for dimensionality reduction, which involves projecting each data point onto only the first few principal components (PCs) in order to obtain lower-dimensional data with the least amount of variance. The first PC is a direction that lowers the predicted data variance. The ith PC minimizes the variance of the projected data by being the inverse of the first $i - 1$ PC.

The primary components of the data covariance matrix can be proven to be eigenvectors. As a result, Eigen decomposition of the data covariance matrix or singular value decomposition of the data matrix is typically used to extract primary components. PCA, closely related to factor analysis, is the most fundamental of the real eigenvector-based multivariate techniques. On the other hand, factor analysis makes additional domain-specific assumptions about the underlying structure and solves matrix eigenvectors. Canonical correlation analysis (CCA) is also tied to PCA. PCA suggests a new orthogonal coordinate system for defining variance in a single dataset, whereas CCA proposes coordinate systems for describing cross-covariance across two datasets.

The purpose of the research is to explore and suggest a suitable type of PCA to reduce the data dimensions, which helps to identify the malware data points significantly.

2. Concept of PCA

PCA attempts to project high-dimensional data onto the feasible smallest-dimensional space. PCA takes into account the variance of each character because a high attribute indicates good class separation, and so minimizes dimensionality. Image processing, movie recommendation systems, and optimizing power distribution across numerous communication channels are some of PCA real-world use cases. It preserves the essential variables while rejecting the less important ones because it is a feature extraction approach [2].

The mathematical concepts employed in the PCA are:

- Variance and covariance
- Eigenvalues and eigenvectors

2.1 Common terms used in PCA

The following are the standard terms widely used in the PCA are discussed below:

Dimensionality: The dimensionality of a dataset refers to the number of characteristics or variables in it. It refers to the total number of columns in a dataset.

Correlation: It expresses the degree to which two variables are intertwined. For example, when one variable is changed, the other variable is also changed. The correlation value can be anywhere between −1 and +1. If the variables are inversely proportional to each other, the result is −1, and if they are directly proportional to each other, the result is +1.

Orthogonal: As a result, the variables have no relationship with one another, and their correlation is zero.

Eigen vectors: If Av is the scalar multiple of v and one has a non-zero vector v and a square matrix M, then v is an eigenvector.

Eigen values: An Eigenvalue is a number that indicates the variance in a specific direction.

Variance: A variance is used to calculate the fluctuation of data points dispersed over the multidimensional graph. In mathematics, it is the average squared deviation from the mean value. The following formula is used to determine $\mathrm{Var}(x)$:

$$\mathrm{Var}(x) = \frac{\sum \left(X_{i} - \overline{X}\right)^{2}}{N} \tag{1}$$

Covariance: Covariance can determine the degree to which comparable components from two sets of grouped data move in the same direction. It is used to uncover relationships and correlations between dataset attributes in layman's terms. The following is the formula for calculating the Cov (x, y):

$$\mathrm{Cov}\ (x) = \frac{\sum \left(X_i - \overline{X}\right) \left(Y_i - \overline{Y}\right)}{N} \tag{2}$$

Covariance matrix: The covariance matrix shows how two variables are related.

Principal components: The new set of data variables created from the original dataset is referred to as PCs. The newly created data variables are pretty valuable and self-contained. They have access to all of the essential data from the original variables.

3. Principal components algorithm

As previously said, the PCs are the converted new characteristics or the result of PCA. The total number of PCs in the dataset is equal or fewer than the total number of original features. Some of the features of these primary components are as follows:

- The significant component must be a linear combination of the original features, and these components must be orthogonal, meaning there must be no link between the two variables.

- As the number of components increases from 1 to n, their relevance decreases, showing that the PC1 is the most essential and the PCn is the least important.

3.1 Steps involved in the PCA algorithm

To carry out the process of PCA the following are the five significant steps to be followed [3]:

I. Standardization

II. Covariance matrix computation

III. Computation of eigenvectors and eigenvalues of the covariance matrix to identify the PCs

IV. Feature vector creation

V. Recast the data along the axes of the PC

3.1.1 Standardization

This step normalizes a set of continuous beginning variables so that their effects on the analysis are consistent.

Standardization is essential before PCA since it is sensitive to the variances of the original variables. Suppose the initial variable ranges differ significantly. In that case, the variables with more comprehensive ranges will outnumber those with smaller ranges (for example, a variable ranging from 0 to 100 will outnumber a variable ranging from 0 to 1), resulting in a skewed outcome. As a result, converting the data to equal scales could be a possible solution to this issue. Subtracting the mean and dividing by the standard deviation for each variable value can be done numerically.

$$Z = \frac{\text{Value} - \text{mean}}{\text{Standard deviation}} \tag{3}$$

After the standardization is complete, all variables will be changed to the same scale.

3.1.2 Covariance matrix computation

The goal of this step is to determine how the variables in the input dataset differ from the mean with each other and whether there is a link between them. Because the variables may become so intertwined that they contain redundant information, the covariance matrix is constructed to find these correlations.

The covariance matrix, a symmetric matrix with $\boldsymbol{p} \times \boldsymbol{p}$ entries, contains all possible pairs of starting variables and their covariances (where p is the number of dimensions). The covariance matrix for a three-dimensional dataset with three variables x, y, and z is a 3×3 matrix of the form:

$$\begin{matrix} \mathbf{Cov}(\boldsymbol{x},\boldsymbol{x}) & \mathbf{Cov}(\boldsymbol{x},\boldsymbol{y}) & \mathbf{Cov}(\boldsymbol{x},\boldsymbol{z}) \\ \mathbf{Cov}(\boldsymbol{y},\boldsymbol{x}) & \mathbf{Cov}(\boldsymbol{y},\boldsymbol{y}) & \mathbf{Cov}(\boldsymbol{y},\boldsymbol{z}) \\ \mathbf{Cov}(\boldsymbol{z},\boldsymbol{x}) & \mathbf{Cov}(\boldsymbol{z},\boldsymbol{y}) & \mathbf{Cov}(\boldsymbol{z},\boldsymbol{z}) \end{matrix} \tag{4}$$

The variances of each starting variable are shown on the main diagonal (top left to bottom right) since a variable covariance with itself equals its variance (Cov(a, a) = Var(a)). The entries of the covariance matrix are symmetric about the principal diagonal because covariance is commutative (Cov(a,b) = Cov(b,a)). This shows that the triangle's upper and lower triangular parts are equal.

The following are the signs of covariance that are related to correlation:

- If the sign of covariance is positive, the two variables will rise or fall in lockstep (i.e., correlated with each other)
- When the sign of covariance is negative, one variable rises while the other falls (i.e., inversely correlated)

3.1.3 Computation of eigenvectors and eigenvalues of the covariance matrix to identify the principal components

In order to uncover the underlying components of the data, eigenvectors and eigenvalues are linear algebra concepts that must be computed from the covariance matrix. Before go into the details of these themes, let us establish what "principal components" mean.

PCs are new variables created by merging or linearly combining essential variables. The new variables (i.e., primary components) are uncorrelated due to these combinations, and the majority of the information from the initial variables is squeezed or compressed into the first components. So, 10-dimensional data provides ten primary components. However, PCA seeks to place as much information as possible in the first component, then as little information in the second, and so on, until the result looks like **Figure 1** below.

One can minimize dimensionality without granting too much information by splitting data into critical components and eliminating components with insufficient data. The remaining components can be regarded as new variables. Because the essential components are produced as linear combinations of the original variables, they are less interpretable and have no significant relevance.

The data directions that explain the most variance, or the lines that include the most data information, are considered essential components in geometric terms. In

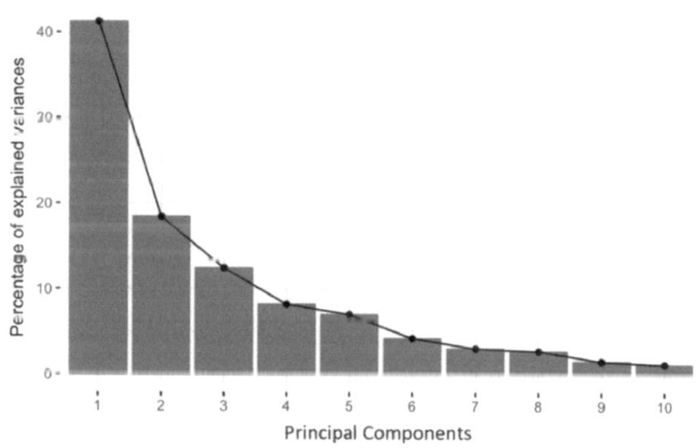

Figure 1.
Principal components vs. percentage of explained variances.

this case, the higher a line variance, the greater the dispersion of data points, and the greater the dispersion along a line, the more information it retains. Put another way; consider the essential components as additional dimensions that provide the proper viewpoint for perceiving and processing data, making it easier to spot differences between observations.

3.1.3.1 Constructing principal components with PCA

Because there are as many variables in the data as there are PCs, the first PC is designed to provide the possible variance in the dataset.

The second major component is determined in the same fashion as the first, except it must be uncorrelated (i.e., parallel to) and account for the following most significant variance. This technique is repeated until the number of variables equals the number of essential components.

3.1.3.2 Finding Eigen values and Eigen vectors

After one has determined the essential components, let us discuss eigenvalues and eigenvectors. Remember that eigenvalues and eigenvectors are always obtained in pairs, with one eigenvalue per eigenvector. In addition, the number is the same as the number of data dimensions. There are three variables in a three-dimensional dataset. Hence there are three eigenvectors with three corresponding eigenvalues. PCs are the eigenvectors of the covariance matrix, and they are the directions of the axis with the most variation. Eigenvalues are the coefficients associated with eigenvectors, whereas eigenvalues are the coefficients attached to eigenvectors. The significant components are ordered in order of significance by arranging the eigenvectors in order of their eigenvalues, from highest to lowest.

Assume the dataset is two-dimensional, with two variables x and y, and the covariance matrix eigenvectors and eigenvalues are:

$$v1 = \frac{0.6778736}{0.7351785}\lambda_1 = 1.284028 \tag{5}$$

$$v2 = \frac{-0.7351785}{0.6778736}\lambda_2 = 0.04908323 \tag{6}$$

The outcome of sorting the eigenvalues in ascending order is $> \lambda 2$, suggesting that the eigenvector of the first PC is v1 and the eigenvector of the second PC is v2. To find the proportion of variance (information) that each component accounts for, divide each component eigenvalue by the total eigenvalues. In the scenario mentioned above, PC1 and PC2 are responsible for 96 and 4% of the data fluctuation.

3.1.4 Feature vector creation

The key components can be identified in order of importance by computing the eigenvectors and sorting them by their eigenvalues in decreasing order. One must decide whether to preserve all of these components or reject those with low eigenvalues and then use the remaining ones to construct the feature vector–matrix at this phase. As a result, the feature vector is just a matrix with the appropriate components eigenvectors as columns. Because only p eigenvectors (components) are left out of n, the final dataset will only have p dimensions.

Combining both eigenvectors v1 and v2 creates a feature vector, as seen in the example above:

$$\begin{bmatrix} 0.6778736 & -0.7351785 \\ 0.7351785 & 0.6778736 \end{bmatrix} \tag{7}$$

Alternatively, one can omit the less relevant eigenvector v2 and solely utilize v1 to generate a feature vector:

$$\begin{bmatrix} 0.6778736 \\ 0.7351785 \end{bmatrix} \tag{8}$$

By eliminating the eigenvector v2, the final dataset dimensionality will be reduced by one, resulting in a loss of information. The loss will be minimal because v2 only carried 4% of the data, and v1 will keep 96% of the data.

The individual must decide whether to maintain all components or delete those not as significant, like in the previous scenario. Because leaving out less significant components is unnecessary if all one wants to do is explain the data in terms of new uncorrelated variables (PCs) without attempting to reduce dimensionality.

3.1.5 Recast the data along the axes of the principal component

The data from previous phases is unchanged except for standardization; all required is to select the primary components and generate the feature vector; however, the input dataset is always in terms of the original axes (i.e., in terms of the initial variables). The third phase, PCA, shifts data from the original axis to the ones indicated by the significant components using a feature vector constructed from the covariance matrix eigenvectors. This is done by multiplying the original dataset transpose by the feature vector transpose. Therefore,

$$\text{Final dataset} = \text{feature vector}^{T}\text{standardized original dataset}^{T} \tag{9}$$

4. When to apply PCA?

PCA is widely applicable for unsupervised machine learning techniques, which helps to reduce the dimensions of the large data where the dataset does not contain the labelled column. The following are some of the situations where the PCA can be applied [4]:

Case 1: One wants to limit the number of variables but cannot figure out which ones to leave out entirely?
Case 2: One wants to make sure variables are unrelated to one another?
Case 3: One thought, if the independent variables are less interpretable?

4.1 Properties of principal components

Suppose the number of primary variables that make up the PCs is less than or equal to the number of variables or data points. In that case, the PCA is complete. The following are the characteristics of primary components [5]:

- They are a set of primary data variables projected in various directions and have qualities similar to those of the original variables.
- In machine learning and data science, dimensionality reduction is a common technique.
- They are orthogonal.
- As one finds PC one by one, the variance or variation of the PCs reduces. This means the first PC is the most volatile, while the last PC is the least volatile.

4.2 Applications of PCA

PCA has a wide range of applications, including the following:

- Face recognition
- Computer vision
- Image compression
- Bioinformatics
- Hidden pattern recognition
- Exploratory data analysis
- Noise filtering
- Finance, data mining, psychology, etc.

4.3 Pros and cons of principal component analysis

For any technique, there will be both positive and negative phases. Likewise, in PCA, its advantages and limitations are the following [6].

4.3.1 Advantages of principal component analysis

Some of the advantages of PCA are:

- Easy to compute
- Speeds up the performance of machine learning algorithms

- Counteracts the issues of high-dimensional data
- Remove correlated features
- Improves the accuracy of the algorithm
- Reduces overfitting
- Enhance visualization

4.3.2 Limitations of principal component analysis

Some of the limitations of PCA are:

- Independent variables become less interpretable
- Data standardization must perform before PCA
- The trade-off between information loss and dimensionality reduction
- Difficult to evaluate the covariance in an appropriate way
- It is sensitive to scale the features
- PCA is not robust against outliers
- PCA assumes a linear relationship between features

5. Variants of PCA

To overcome the limitations of PCA, there are different types of PCA are available that suit for the appropriate type of data are listed below [7, 8]:

- Normal PCA
- Sparse PCA
- Randomized PCA
- Incremental PCA
- Kernel PCA

5.1 Normal PCA

PCA in Machine learning is applied for unsupervised learning to reduce the dimension of the data from high dimensional space to low dimensional space. The above section discusses the standard normal PCA, which applies to most of the datasets as a default form of PCA using unsupervised learning. To construct any type of PCA especially for normal PCA the above discussed five significant steps are involved for dimensionality reduction [9, 10]. The following sessions briefly discuss the other variants of PCA in machine learning and its characteristics.

5.2 Sparse PCA

One of PCA significant flaws is that the PCs are dense in most circumstances, implying that the majority of the loadings are non-zero. The model is difficult to interpret since each significant component is a linear combination of all the original variables. However, each axis may correspond to a specific gene in machine learning tasks such as gene analytics. In such instances, one can readily analyse the model and comprehend the physical meaning of the loading and the PCs if the majority of the entries in the loadings are zeros. Sparse PCA is a variant of PCA that uses sparse loading to build interpretable models. In Sparse PCA, each PC is a linear combination of a subset of the original variables.

5.3 Randomized PCA

The PCs are estimated using the low-rank matrix approximation in traditional PCA. However, this strategy becomes costly with large datasets and makes the entire process challenging to scale. One can approximate the first K PCs faster than traditional PCA by randomizing how the dataset singular value decomposition occurs.

5.4 Incremental PCA

The above-described PCA variants need the entire training dataset to be stored in memory. Incremental PCA can be employed when the dataset is too huge to fit in memory. It divides the dataset into mini-batches, each of which can fit into memory, and then feed each mini-batch to the incremental PCA algorithm one at a time.

5.5 Kernel PCA

A typical linear technique is PCA. It works well with linearly separable datasets. However, if the dataset contains non-linear relationships, the results will be unfavourable. Kernel PCA is a technique that uses the "kernel trick" to project linearly inseparable data into a higher dimension where it may be separated linearly. Many different kernels are commonly employed, including linear, polynomial, RBF, and sigmoid.

6. Case study on variants of PCA in mobile malware detection

To explore the dimensionality reduction using normal PCA and its variants for mobile malware detection in the CICMalDroid_2020 dataset [11] are experimented. The dataset is taken from the University of New Brunswick (UNB), Canadian Institute of Cyber Security (CIC). The dataset consists of 11,598 records with 471 feature attributes. The dataset commences with 17,341 Android samples gathered from VirusTotal, the Contagio security blog, AMD, MalDozer, and other sources. Samples were obtained between December 2017 and December 2018. Detecting Android apps with malicious data points is crucial for cyber security specialists. There are five key categories in the dataset includes, Adware, Banking malware, SMS malware, Riskware, and Benign are the different forms of malicious software. The experiment is carried out in a Python Jupyter notebook environment using sklearn library [12–16].

6.1 Results of PCA in machine learning for mobile malware detection

The following are the outcomes of normal PCA and variants of PCA where the 471 feature dimensions are reduced into two PCs are visualized below. **Figure 2** shows the importing data into the Python Jupyter notebook environment.

Figure 3 shows the data pre-processing to check whether the data contains any null values or not. Data pre-processing is an important step in the machine learning process, and it helps to purify the irrelevant and undefined raw data into the relevant defined form.

Figure 3 depicts that the dataset does not contain any null values, and it is fit for further processing (i.e.) from the results value '0'indicates there are no null values in the data.

Figure 4 shows the removal of duplicate data values to ensure the originality of the dataset. Duplicate data leads to misinterpretation of the results.

```
df = pd.read_csv("C:/Users/Acer/Desktop/feature_vectors_syscallsbinders_frequency_5_Cat.csv")
#df = pd.read_csv(r"C:\Users\hp\Desktop\Malware dataset_CCT.csv", encoding = 'utf-8')
df.shape

(11598, 471)
```

Figure 2.
Data import—CICMalDroid_2020 dataset.

```
#Data Pre-processing
print(df.isnull().sum())

ACCESS_PERSONAL_INFO___      0
ALTER_PHONE_STATE___         0
ANTI_DEBUG_____              0
CREATE_FOLDER_____           0
CREATE_PROCESS`_____         0
                            ..
watchRotation                0
windowGainedFocus            0
write                        0
writev                       0
Class                        0
Length: 471, dtype: int64
```

Figure 3.
Check if the data contains any null values or not.

Figure 4 depicts that out of 11,598 records, 72 were duplicated, and the duplicated records were dropped. After dropping the 72 duplicate records, now the dataset consists of 11,526 instances with 471 features.

Figure 5 shows the data splitting for training and testing so that the machine learning model can detect and cluster the mobile malware data points in the dataset. Splitting the data for training and testing is a significant phase in the machine learning process. So, the data will be adequately trained and provide the best results in testing, which helps to derive a high efficacy rate.

Figure 5 explains that the dataset is divided into 70% for training and 30% for testing (i.e.) out of 11,526 records, 8068 are used for training and 3458 samples are used for testing the model. Now, the dataset is suitable to perform the PCA with the machine learning technique.

Figure 6 shows the feature scaling; before applying PCA, one must scale the data so that the data can be properly scaled within a particular range appropriately to support data modelling. Without incorporating feature scaling, during the model development the data takes more time to fit into the prescribed model form.

Figure 6, depicts the method for feature scaling using MinMaxScaler and standard scalar to bring the scattered data points within a typical specified range. Hence, the data is further applicable for PCA.

```
#to check for any duplicate data values
df2 = df1.drop_duplicates()
df2.shape

(11526, 471)
```

Figure 4.
Drop duplicate values.

```
#Splitting data for training and testing
X = df2.iloc[:,:-1].values #features
Y = df2['Class'].values #target variable

from sklearn.model_selection import train_test_split

X_train, X_test, y_train, y_test = train_test_split(X, Y, test_size = 0.3,random_state = 0)
print("Training samples: {}; Test samples: {}".format(len(X_train), len(X_test)))

Training samples: 8068; Test samples: 3458
```

Figure 5.
Train and test split.

```
#Feature Scaling
from sklearn.preprocessing import MinMaxScaler
ms = MinMaxScaler()
X = ms.fit_transform(X)

from sklearn.preprocessing import StandardScaler
sc = StandardScaler()
X_train = sc.fit_transform(X_train)
X_test = sc.transform(X_test)
X = sc.fit_transform(X)
```

Figure 6.
Feature scaling.

Figure 7 shows the normal PCA method used in the dataset to reduce the 471 featured dimensions into two PCs. It also shows that from the explained variance PC1 has more information than PC2. Normal PCA is the default form of PCA, more suitable for all kinds of data to reduce the dimensions effectively without any information loss.

Figure 8 shows the normal PCA results of two PCs as PC1 and PC2.

Figure 8 represents that the total 471 features are reduced into two PCs PC1 and PC2, without removing any of the data features. This helps to train the data and develop the machine learning model effectively with less memory consumption.

Figure 9 shows the method of sparse PCA in mobile malware data. Similarly, the sparse PCA is widely suited for sparse data so that the 471 data features are reduced into two set of PCs PC1 and PC2.

Figure 10 shows the method of randomized PCA in mobile malware data. Randomized PCA is suitable for big data processing so that the features are randomly selected to derive the two set of PCs PC1 and PC2.

Figure 11 shows the method of incremental PCA in mobile malware data. Incremental PCA is similar to randomized PCA, but it gradually increases the batch size to reduce the total number of features into two set of PCs PC1 and PC2.

Figure 12 shows the method of kernel PCA in mobile malware data. Kernel PCA is widely applicable for non-linear data modelling. It also helps to reduce the dimensions of the data based on the kernel function like gamma etc. to derive the two or more sets of PCs PC1and PC2.

In this case study, the 471 data features of CICMalDroid_2020 dataset is transformed into two set of PCs PC1 and PC2 for all variants of PCA, depending upon the suitability of the data values.

```
# Applying PCA function on training
# and testing set of X component
from sklearn.decomposition import PCA

pca = PCA(n_components = 2)

X_train = pca.fit_transform(X_train)
X_test = pca.transform(X_test)

explained_variance = pca.explained_variance_ratio_
print(explained_variance)

[0.70241931 0.29758069]
```

Figure 7.
Normal PCA.

	Principal component 1	Principal component 2
8113	-3.371104	0.361204
8114	-2.592353	-0.461123
8115	3.150280	1.218977
8116	8.846860	5.503562
8117	-2.846843	-1.267918

Figure 8.
Normal PCA with PC1 and PC2.

```
#Sparse PCA
from sklearn.decomposition import SparsePCA

spca = SparsePCA(n_components=2, alpha=0.0001)
X_train = spca.fit_transform(X_train)
X_test = spca.transform(X_test)

#explained_variance = spca.explained_variance_ratio_
#print(explained_variance)
```

```
principal_data_Df = pd.DataFrame(data = X_train
             , columns = ['Principal component 1', 'Principal component 2'])
principal_data_Df
```

	Principal component 1	Principal component 2
0	-24.650157	13.666278
1	-33.683180	17.554447
2	-0.555096	-4.540046
3	2.199703	0.124262
4	3.263998	-0.028142
...	...	...
8063	6.213232	9.590552
8064	2.451784	0.380245
8065	-4.936274	-9.065537
8066	-2.058104	-1.706935
8067	1.720193	0.536436

8068 rows × 2 columns

Figure 9.
Sparse PCA.

```
#Randomized PCA
from sklearn.decomposition import PCA

rpca = PCA(n_components=2, svd_solver='randomized')
X_train = rpca.fit_transform(X_train)
X_test = rpca.transform(X_test)

explained_variance = rpca.explained_variance_ratio_
print(explained_variance)
```

```
[0.70241931 0.29758069]
```

```
principal_data_Df = pd.DataFrame(data = X_train
             , columns = ['Principal component 1', 'Principal component 2'])
principal_data_Df
```

	Principal component 1	Principal component 2
0	-2.783878	-0.664049
1	-2.187618	-0.819019
2	-0.833045	0.270127
3	23.384158	-12.316224
4	-3.639008	-2.031365
...	...	...
8113	-3.207486	0.341588
8114	-2.466535	-0.438440
8115	2.997384	1.160743
8116	8.417485	5.237434
8117	-2.708675	-1.205984

8118 rows × 2 columns

Figure 10.
Randomized PCA.

```
#Incremental PCA
from sklearn.decomposition import IncrementalPCA

ipca = IncrementalPCA(n_components=2, batch_size=200)
X_train = ipca.fit_transform(X_train)
X_test = ipca.transform(X_test)

explained_variance = ipca.explained_variance_ratio_
print(explained_variance)
```

```
[0.70241931 0.29758069]
```

```
principal_data_Df = pd.DataFrame(data = X_train
            , columns = ['Principal component 1', 'Principal component 2'])
principal_data_Df
```

	Principal component 1	Principal component 2
0	-2.783878	-0.664049
1	-2.187618	-0.819019
2	-0.833045	0.270127
3	23.384158	-12.316224
4	-3.639008	-2.031355
...	...	...
8113	-3.207486	0.341588
8114	-2.466535	-0.438440
8115	2.997384	1.160743
8116	8.417485	5.237434
8117	-2.708675	-1.205984

8118 rows × 2 columns

Figure 11.
Incremental PCA.

```
#Kernel PCA
kpca = KernelPCA(kernel='rbf', gamma=1)
X_kpca = kpca.fit_transform(X_train)

plt.title("Kernel PCA")
plt.scatter(X_kpca[:,0], X_kpca[:,1], c=y_train)
plt.show()
```

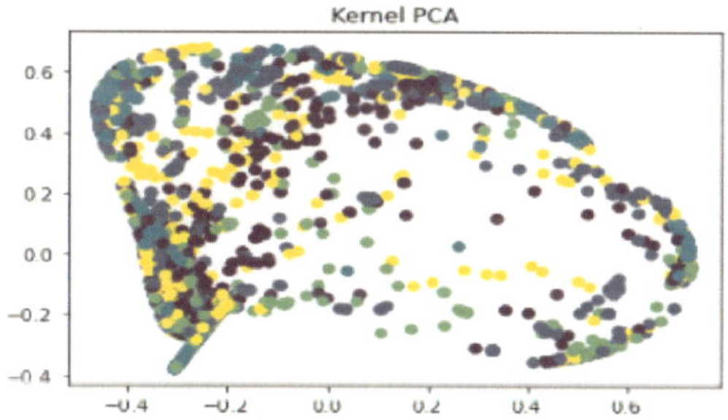

Figure 12.
Kernel PCA.

6.2 Observations

Based on the results obtained from the variants of PCA for mobile malware detection depicts that for the given CICMalDroid_2020 dataset is discussed. It is a numerical labelled data that highly supports normal standard PCA technique. It helps to reduce the dimensions of the PCA from 471 features into two sets of PCs (PC1 and PC2). It supports to processing the data model quickly and forms the clusters

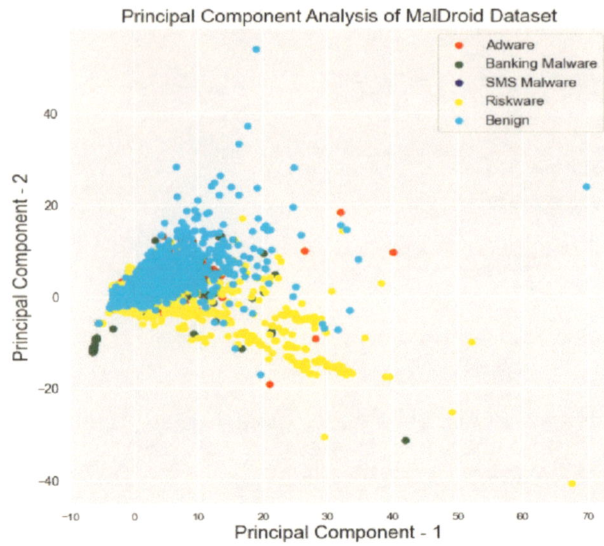

Figure 13.
Class of malwares in the CICMalDroid_2020 dataset.

Class	Type of malware	Size
1	Adware	1253
2	Banking malware	2100
3	SMS malware	3904
4	Riskware	2546
5	Benign	1795

Table 1.
CICMalDroid_2020 dataset—sample size

effectively. **Figure 13** shows the group of clusters that discovered the five different malware involved in the CICMalDroid_2020 dataset based on PC1containing huge information about the dataset of normal PCA.

Table 1 shows the size of malware samples available under the category of Adware, Banking malware, SMS malware, Riskware, and Benign.

Hence, the other variants of PCA, such as sparse PCA are applicable for sparse data, randomized PCA and incremental PCA are suitable for big data processing and kernel PCA is widely supported by non-linear data modelling. So, depending upon the type of data and their accessibility, the appropriate type of PCA is incorporated into machine learning algorithms specifically for unsupervised learning for dimensionality reduction to develop a suitable predictive model. It also helps to identify the hidden patterns of the data effectively.

7. Conclusion

This book chapter on "Evaluation of PCA Variants to assess their suitability for Mobile Malware Detection" describes briefly the concept of PCA, the common terminologies involved in PCA, PCs in PCA, mathematical properties of PCA, steps involved in PCA algorithm, explanation about the process of developing the PCA model in a machine learning perspective, applications of PCA, advantages and limitations of PCA. Different variants of PCA are experimented with a small case study by exploring the various type of PCA. This helps to find out the suitable variant of PCA for mobile malware detection in the CICMalDroid_2020 dataset based on the machine learning framework. As an outcome, for the given dataset, the normal standard PCA provides the appropriate results for the PCs to discover the malware data points accurately. Thus, this chapter will be a ready reckoner for the learners to know about the concept of PCA in machine learning, and its variants suitable for mobile malware detection are discussed in detail.

Acknowledgements

This work is supported by the DST-CURIE-AI project during 2021–2023 by the Centre for Cyber Intelligence established under the Centre for Machine Learning and Intelligence, Avinashilingam Institute for Home Science and Higher Education for Women, Coimbatore-43, Tamilnadu, India.

Conflict of interest

The authors declare no conflict of interest.

Nomenclature

PCA	Principal component analysis
CCA	Canonical correlation analysis
CIC	Canadian Institute of Cyber Security
Var	Variance
Cov	Covariance
UNB	University of New Brunswick
PC	Principal components
X, Y, Z	Variables
RBF	Radial basis function
n	Number of features
m	Number of PCs
v	Eigenvector
λ	Eigenvalue

Author details

Padmavathi Ganapathi[1*], Shanmugapriya Dhathathri[2] and Roshni Arumugam[3]

1 Department of Computer Science, Avinashilingam Institute for Home Science and Higher Education for Women, Coimbatore, Tamilnadu, India

2 Department of Information Technology, Avinashilingam Institute for Home Science and Higher Education for Women, Coimbatore, Tamilnadu, India

3 Centre for Cyber Intelligence, Avinashilingam Institute for Home Science and Higher Education for Women, Coimbatore, Tamilnadu, India

*Address all correspondence to: padmavathi_cs@avinuty.ac.in

References

[1] Available from: https://www.sartorius.com/en/knowledge/science-snippets/what-is-principal-component-analysis-pca-and-how-it-is-used-507186. [Accessed: 26 April 2022]

[2] Available from: https://en.wikipedia.org/wiki/Principal_component_analysis. [Accessed: 26 April 2022]

[3] Available from: https://builtin.com/data-science/step-step-explanation-principal-component-analysis. [Accessed: 26 April 2022]

[4] Available from: https://www.javatpoint.com/principal-component-analysis [Accessed: 06 March 2022]

[5] Available from: https://intellipaat.com/blog/a-brief-introduction-to-principal-component-analysis/. [Accessed: 26 April 2022]

[6] Available from: https://www.i2tutorials.com/what-are-the-pros-and-cons-of-the-pca/. [Accessed: 27 April 2022]

[7] Available from: https://www.keboola.com/blog/pca-machine-learning. [Accessed: 27 April 2022]

[8] Available from: https://aiaspirant.com/types-of-pca/. [Accessed: 27 April 2022]

[9] Available from: https://www.analyticssteps.com/blogs/introduction-principal-component-analysis-machine-learning. [Accessed: 27 April 2022]

[10] Available from: https://intellipaat.com/blog/a-brief-introduction-to-principal-component-analysis/. [Accessed: 27 April 2022]

[11] Available from: https://www.unb.ca/cic/datasets/maldroid-2020.html. [Accessed: 06 March 2022]

[12] Benahmed L, Houichi L. The effect of simple imputations based on four variants of PCA methods on the quantiles of annual rainfall data. Environmental Monitoring and Assessment. 2018;**190**(10):1-14. DOI: 10.1007/s10661-018-6913-y

[13] Wang Z, Han D, Li M, Liu H, Cui M. The abnormal traffic detection scheme based on PCA and SSH. Connection Science. 2022;**34**(1):1201-1220. DOI: 10.1080/09540091.2022.2051434

[14] Manzano C, Meneses C, Leger P, Fukuda H. An empirical evaluation of supervised learning methods for network malware identification based on feature selection. Complexity. 2022; **2022**:1-18. DOI: 10.1155/2022/6760920

[15] Aurangzeb S, Anwar H, Naeem MA, Aleem M. BigRC-EML: Big-data based ransomware classification using ensemble machine learning. Cluster Computing. 2022. DOI: 10.1007/s10586-022-03569-4

[16] Rajadurai H, Gandhi UD. An empirical model in intrusion detection systems using principal component analysis and deep learning models. Computational Intelligence. 2020;**37**(3):1111-1124. DOI: 10.1111/coin.12342

Chapter 4

Principal Component Analysis and Artificial Intelligence Approaches for Solar Photovoltaic Power Forecasting

Souhaila Chahboun and Mohamed Maaroufi

Abstract

In recent years, renewable energy sources have experienced remarkable growth. However, their spatial and temporal diversity makes their large-scale integration into the current power grids difficult, as the balance between the electricity output and the consumption must be maintained at all times. Therefore, it is important to focus on the resources forecast to enhance the integration of renewable energy sources, such as solar in this study. In this article, a comparative analysis of two main machine learning methods was conducted for the prediction of the hourly photovoltaic output power. Furthermore, since various factors, such as climate variables, can impact the solar photovoltaic power and complicate the prediction process, the principal component analysis was employed to investigate the interactions between the multiple predictors and minimize the dimensionality of the datasets. The prevalent factors were then used in the predictive models as inputs. This field research is very crucial because the higher the prediction accuracy, the greater the profit for energy dealers and the lower the costs for customers.

Keywords: photovoltaic power, machine learning, principal component analysis, prediction

1. Introduction

The primary driver of the economic progress of a country is energy [1]. Recently, renewable energy sources have become increasingly popular. Solar energy is gaining popularity due to its low pollution, great energy efficiency, and adaptability [2].

However, the output power of solar energy is strongly impacted by weather and other environmental factors, restricting its deployment on a broad scale. In the solar power generating system, research on photovoltaic (PV) power generation prediction is consistently one of the most prominent topics of study [3].

The most widely employed a physical model of forecasting is numerical weather prediction. The numerical weather forecast model is computationally complex due to the fluctuation and unpredictable character of the atmosphere. Therefore, as the area of computer science expands and its ability to deal with non-linearity improves, machine learning offers a prospective advantage for renewable energy forecasting. The precision of the input data and the machine learning techniques employed determine the efficiency of the predictive models [4]. Moreover, even if the input–output data connection is complex, machine learning methods use historical data sets to construct a relationship between them. As a result, it is essential to use appropriate data to address the problem efficiently [5].

In recent years, a growing number of algorithms have been employed in the field of PV prediction, resulting in ever-improving forecast accuracy. The present state of PV forecasting techniques can be mainly summed up in Neural Network, Multivariate Adaptive Regression Splines, Boosting, Bagging, K-nearest-neighbor etc. However, the large number of variables and irrelevant or redundant information can make forecasting difficult, necessitating a large amount of computer power and resulting in inefficient and erroneous results. Feature reduction approaches are presented as a solution to overcome this challenge [6].

This approach was adopted by a number of researchers. For instance, Souhaila et al. [7] carried out a principal component analysis (PCA) to decrease the number of interconnected variables. These dominant factors were then employed in the predictive models as inputs. Qijun et al. [2] employed both PCA and Support Vector Machine for PV power prediction. Malvoni et al. [8, 9] created a PV forecast model based on a hybrid PCA– Least-squares support vector machine (LSSVM).

Given the challenges, mentioned above, related to the field of PV power prediction, the aim of this study is to determine the most effective data and machine learning algorithms for accurate PV power output forecast. Moreover, this study investigates the impact of data pre-processing approaches, mainly Yeo-Johnson transformation (YJT), correlation analysis, and PCA technique, on machine learning prediction accuracy. The two main machine learning algorithms used in this study are Multiple Linear Regression and Cubist Regression Finally, the most common error metrics and residual analysis were used to assess the accuracy of the predictive models.

2. Data preparation

Data preparation is necessary to get the best results from machine learning algorithms. Some machine learning algorithms require data to be in a specific format. As a result, it is vital to arrange the data so that various machine learning algorithms have the best chance of solving the studied problem. In our case, two techniques were employed for data preparation namely Yeo-Johnson transformation (YJT) and correlation analysis.

2.1 Data source

In this study, we used the PV power data from a PV power platform in Morocco, having a total capacity of 6 KW. For the input data, we made advantage of a free data source that gives solar energy and meteorological information. The inputs used in our forecasting models are presented in **Tables 1–3**:

2.2 Yeo-Johnson transformation

In general, many data include variables with a non-normal distribution (gaussian). However, they are frequently skewed in their distributions. Preprocessing the

Parameter	Unit	Symbol
Top of Atmosphere radiation	$Wh/_{m^2}$	TOA
Global Horizontal irradiation	$Wh/_{m^2}$	GHI
Beam Horizontal irradiation	$Wh/_{m^2}$	BHI
Diffuse Horizontal irradiation	$Wh/_{m^2}$	DHI
Beam Normal irradiation	$Wh/_{m^2}$	BNI

Table 1.
Solar irradiation data.

Parameter	Unit	Symbol
Relative Humidity	%	RH
Wind Speed	$m/_s$	WS
Ambient Temperature	°C	Tamb
Pressure	hPa	P

Table 2.
Meteorological data.

Parameter	Unit	Symbol
Module Temperature	°C	Tm
Efficiency	%	Eff
Month	—	Month
Day	—	Day
Hour	—	Hour

Table 3.
Supplemental data.

variables to make them more normal is common when dealing with such data. The Box-Cox and Yeo-Johnson transformations (YJT) are two well-known methods for this. Yeo and Johnson (2000) improved the Box-Cox transformation to create a one-parameter family that can transform both positive and negative variables [10]. YJT is defined by Eq. (1):

$$y^{(\lambda)} = \begin{cases} \dfrac{(y+1)^{\lambda}-1}{\lambda} & \lambda \neq 0 \; and \; y \geq 0 \\ \ln(y+1) & \lambda = 0 \; and \; y \geq 0 \\ \dfrac{-\left((-y+1)^{2-\lambda}-1\right)}{2-\lambda} & \lambda \neq 2 \; and \; y < 0 \\ \ln(-y+1) & \lambda = 2 \; and \; y < 0 \end{cases} \quad (1)$$

This transformation is ideal for correcting left and right skew when $\lambda > 1$ and $\lambda < 1$ respectively, whereas when $\lambda = 1$, the linear connection is established.

3. Materials and methods

3.1 Correlation analysis

The correlation between the parameters of the model has a significant impact on the accuracy of the forecasted models. To simplify computations, the correlation of different inputs with PV power generation was evaluated. The correlation matrix is calculated with the help of the covariance Eq. (2) and correlation metrics Eq. (3). Below are the equations:

$$cov(a,b) = \frac{1}{N}\sum_{i=1}^{N}(x_i - \overline{x}) \times (y_i - \overline{y}) \tag{2}$$

$$corr(a,b) = \frac{cov(x,y)}{s(x) \times s(y)} \tag{3}$$

where $\overline{x}, \overline{y}$ represent the means of the x and y values, respectively, and s represents the standard deviation. It's used to figure out how dispersed the data is around the mean value.

3.2 Principal component analysis

The dataset must be pre-processed and dimensionally reduced before the training of the machine learning models. Principal component analysis (PCA) is a dimensionality reduction and feature extraction technique based on linear transformations. Using an orthogonal transformation, this approach converts correlated variables into mutually uncorrelated variables. The major components calculated from the Eigen vector of the covariance matrix can be lower or equal to the original variables. The first principal components, which reflect a high correlation between input variables, account for the majority of the variance [11].

3.3 Forecasting models

In this study, we decided to assess the efficiency of two popular machine learning methods using the R software [12].

3.3.1 Multiple linear regression

Multiple Linear Regression (MLR) is a technique for predicting the power generated by solar PV panels using a range of predictor variables. The following is the regression equation (see Eq. (4)):

$$Y = \beta_0 + \beta_1 X_1 + \beta_2 X_2 \dots .. + \beta_k X_k \tag{4}$$

where $X_1.X_2, \dots, X_n$ are predictor variables and $\beta_1,\beta_2, \dots \beta_n$ are their coefficients.

3.3.2 Cubist regression

Cubist (CB) is a rule-based approach that uses building rules to generate regression solutions. A rule is generated for each leaf in a regression tree, and it is linked to the data it contains. The linear combination of rules that occurs when all rules are constructed is used to make final predictions [13]. The CB model incorporates boosting with training committees, which is comparable to the approach of boosting by generating a sequence of trees with changed weights successively. The number of neighbors of the CB model is used to modify the rule-based prediction. The models created by two linear models in the CB model are written as follows in Eq. (5), [14]:

$$\hat{y}_{par} = (1 - a) \times \hat{y}_p + a \times \hat{y}_c \tag{5}$$

where $\hat{y}_c$ is the forecast of the current model and $\hat{y}_p$ is the prediction of the parent model.

3.4 Error metrics

We randomly divided the data into a training set and a testing one to evaluate the investigated models and measure their prediction power. Eqs. (6)–(8) establish the error metrics used to assess the accuracy of the predictive models.

$$R^2 = 1 - \frac{\sum_{i=1}^{n}\left(Pout_i - \widehat{Pout_i}\right)^2}{\sum_{i=1}^{n}\left(Pout_i - \overline{Pout}\right)^2} \tag{6}$$

$$\text{RMSE} = \sqrt{\frac{1}{n}\sum_{i=1}^{n}\left(Pout_i - \widehat{Pout_i}\right)^2} \tag{7}$$

$$\text{MAE} = \frac{1}{n}\sum_{i=1}^{n}\left|Pout_i - \widehat{Pout_i}\right| \tag{8}$$

4. Results

4.1 Correlation analysis results

A correlation study was performed, as previously indicated, to check the connection between the input variables and the output power, thereby selecting the closely related factor parameters that should be kept as inputs to the prediction models (see **Figure 1**).

4.2 Principal component analysis results

As previously explained, PCA was used to determine the most essential data variables to be used in the training of the machine learning models. The variance distribution of the principal components (PCs) (PC1–PC9) is depicted in the Scree plot in **Figure 2**. According to the eigenvalues, the cumulative variance of PC1 through PC3 is **90.4%**. As a result, the first three major components were recognized as the primary model inputs and were sufficient for the development of our predictive models.

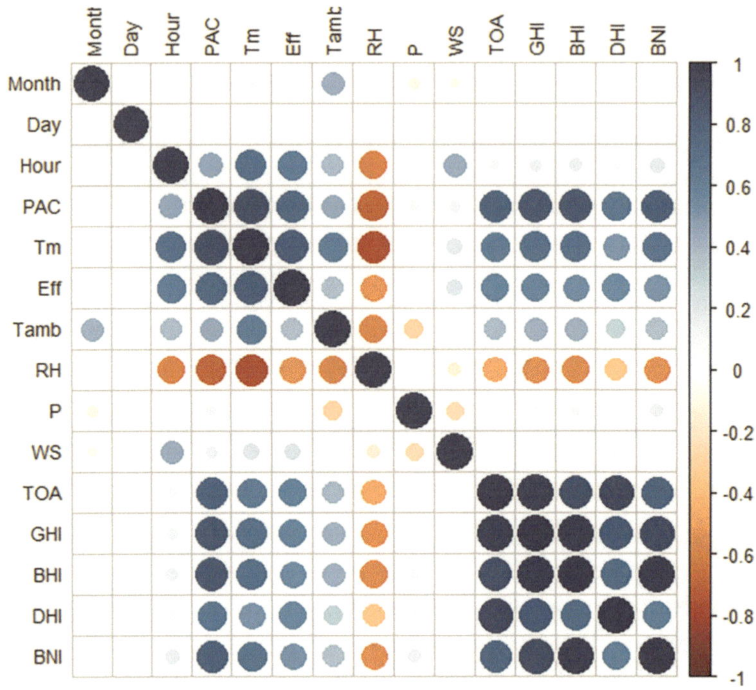

Figure 1.
Correlation matrix.

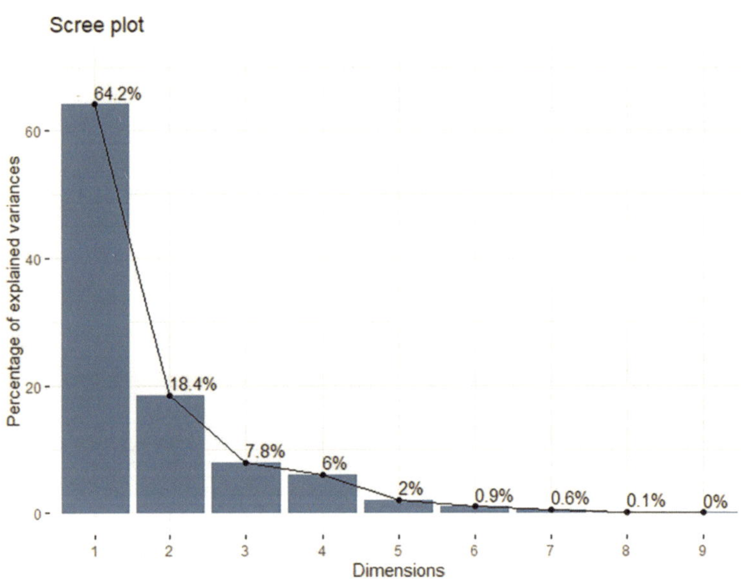

Figure 2.
Scree plot.

The main variables of each of the PCs were selected from the top three variables in **Table 4** with a value greater than **0.60** [15]. **GHI, BHI,** and **BNI** were selected for PC1. For PC2, **Hour, Tm,** and **Eff** were identified. Finally, only **Tamb** was chosen for PC3.

Factor	PC1	PC2	PC3
Hour	0.01	**0.98**	0.16
Tm	0.47	**0.68**	0.33
Eff	0.28	**0.60**	0.10
Tamb	0.19	0.24	**0.94**
TOA	0.57	0.07	0.15
GHI	**0.76**	0.10	0.17
BHI	**0.88**	0.11	0.18
DHI	0.34	0.08	0.10
BNI	**0.94**	0.14	0.13

Table 4.
PCA results.

Algorithm	Raw data			Reduced data (PCA)		
	R^2	RMSE (KW)	MA (KW)	R^2	RMSE (KW)	MAE (KW)
MLR	0.9016	0.6642	0.5036	0.9147	0.7894	0.6127
CB	0.9944	0.1575	0.1032	0.9914	0.2499	0.1597

Table 5.
Performance metrics results—Training phase 80%.

Algorithm	Raw data			Reduced data (PCA)		
	R^2	RMSE (KW)	MAE (KW)	R^2	RMSE (KW)	MAE (KW)
MLR	0.8963	0.6780	0.5155	0.9218	0.7578	0.5922
CB	0.9807	0.2921	0.1830	0.9821	0.3622	0.2191

Table 6.
Performance metrics results—Testing phase 20%.

4.3 Performance metrics

Tables 5 and **6** show the forecast performance results in the case of raw data and reduced data resulting from PCA method.

Scatter plots (see **Figure 3**) reveal more information about the model's effectiveness. All points in a good model should be close to the diagonal line and have no practical dependencies.

4.4 Residual analysis

The difference between the actual and expected values is known as residual. The Residual vs. fitted values plot is the first plot in our residual analysis (see **Figure 4**). It is one of the most used model validation graphs. This figure detects outliers and error

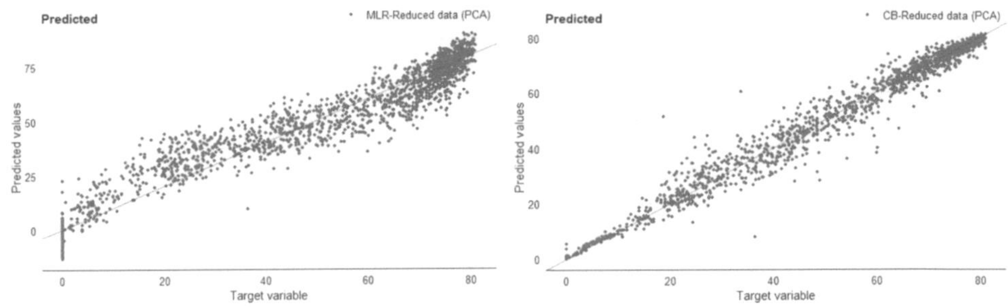

Figure 3.
Predicted versus observed values plots.

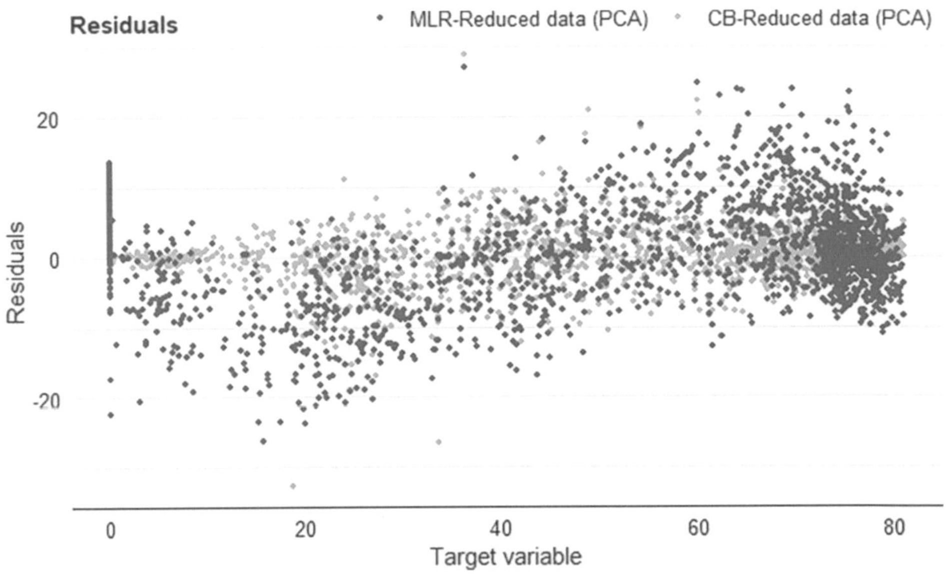

Figure 4.
Residuals versus observed values plot.

dependencies. The precision of the forecast for that particular value is shown by the distance from the x-axis (0 line).

Moreover, the Residual density plot, as shown in **Figure 5**, can be very informative. If the majority of the residuals are not grouped at zero, the model outputs will likely be biased.

Finally, the last plot (**Figure 6**) is the residual boxplot. It depicts the distribution of absolute residual values.

5. Discussion

Based on the results of the correlation analysis (see **Figure 1**), month, day, WS, and P variables have the lowest correlation with the PV output power, whereas solar irradiations and Tm have the strongest correlation with the PV power. Furthermore, all of the variables have a negative correlation with RH parameter. As RH rises, the PV power decreases. Moreover, the relationship between Tamb, Hour, Eff, and PV output

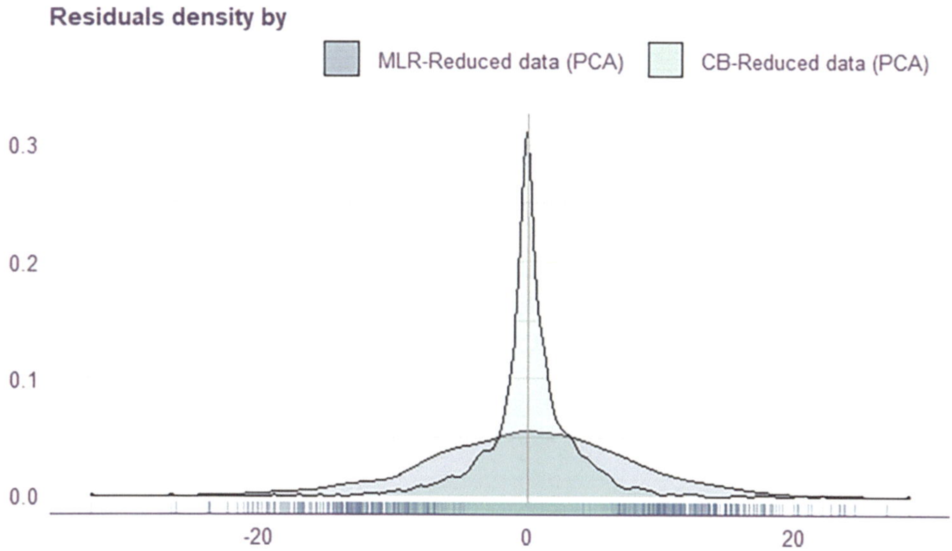

Figure 5.
Residual density.

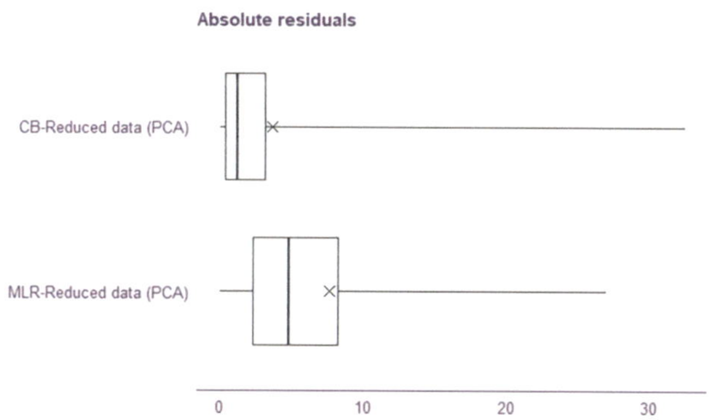

Figure 6.
Residual boxplot.

power appears to be neither strong nor weak. As a result, we simplified the PV power forecast method by removing the variables Month, Day, RH, P, and WS from the input data and keeping other variables as the main inputs to our regression models.

The PCA method showed three major factor components that influence PV power and reach up to 90.4% of the total variable variance. As a result, the PCA technique was used to identify the most significant variables, which are then used in the proposed models.

The results of performance metrics, on the other hand, in **Tables** 5 and 6, the CB technique provided the best balance between the forecasted and observed values, with an R^2 = 98.21% in the testing phase and R^2 = 99.14% in the training one. This is owing to the fact that linear models lose accuracy when the dependencies are not linear, as is the case with solar PV output. Moreover, by comparing the results obtained in the case of raw data and reduced data resulting from the PCA analysis, the results are clearly superior, demonstrating the critical importance of this dimensionality reduction approach, which allows for cost and efficiency savings.

Moreover, the **Figure 3** gives extra information on model efficiency in addition to the error metrics presented above. All observed points should, in theory, be close to the diagonal line, which is the case of the CB algorithm.

Finally, several plots have been presented above to help in the analysis of the predictive models in terms of residuals. From the plot of residual vs. observed values presented in **Figure 4**, the CB method obviously surpasses the MLR method in terms of prediction accuracy, since residuals in CB are more localized around the x-axis than in MLR.

In addition, compared to MLR, **Figure 5** shows that residuals in CB are more localized around zero. Furthermore, looking at the Residual boxplots in **Figure 6**, we can see that CB has the smallest number of residuals compared to MLR, which has a much larger range of residuals.

All the results obtained show the superiority of the CB algorithm in predicting the PV power compared to the classical approach MLR.

6. Conclusions

In the sector of PV power forecasting, machine learning techniques within artificial intelligence offer a lot of potential. The main benefit of these approaches is their ability to handle complex problems and take into consideration a large number of input factors, However, it is worth noting that selecting an optimum number of input variables is beneficial for successful machine learning, since large datasets can be difficult to analyze and interpret. As a result, the PCA approach is critical, as it allows for faster computations and storage space savings, as well as the removal of redundant variables, multicollinearity, and noise.

Finally, the comparison of machine learning approaches for PV power forecasting will aid energy suppliers in identifying the best algorithms for effectively and safely handling PV-integrated power.

Nomenclature

BHI	beam horizontal irradiation
BNI	beam normal irradiation
CB	cubist
DHI	diffuse horizontal irradiation
Eff	efficiency
GHI	global horizontal irradiation
MAE	mean absolute error
MLR	multiple linear regression
P	pressure
PCA	principal component analysis
PV	photovoltaic
RH	relative humidity
RMSE	root mean square
R^2	R-squared
Tamb	ambient temperature
Tm	module temperature

TOA	top of atmosphere radiation
WS	wind speed
YJT	Yeo-Johnson transformation

Author details

Souhaila Chahboun* and Mohamed Maaroufi*
Mohammed V University in Rabat, Mohammadia School of Engineers, Rabat, Morocco

*Address all correspondence to: souhaila_chahboun@um5.ac.ma and maaroufi@emi.ac.ma

References

[1] Chahboun S, Maaroufi M. Novel comparison of machine learning techniques for predicting photovoltaic output power. International Journal of Renewable Energy Research. 2021;**11**(3): 1205-1214

[2] Qijun S, Fen L, Jialin Q, Jinbin Z, Zhenghong C. Photovoltaic power prediction based on principal component analysis and support vector machine. 2016 IEEE Innovative Smart Grid Technologies - Asia (ISGT-Asia); 2016; 815-820. DOI: 10.1109/ISGT-Asia.2016.7796490

[3] Souhaila C, Mohamed M. Ensemble methods comparison to predict the power produced by photovoltaic panels. Procedia Computer Science. 2021;**191**: 385-390. DOI: 10.1016/j.procs.2021.07.049

[4] Moslehi S, Reddy TA, Katipamula S. Evaluation of data-driven models for predicting solar photovoltaics power output. Energy. 2018;**142**:1057-1065

[5] Wu Y, Wu M, Bao L, Li C. Short-term power forecasting of photovoltaic power generation based on similar day and improved principal component analysis. Journal of Computers. 2020;**31**(5):187-197

[6] Ziane A, Necaibia A, Sahouane N, Dabou R, Mostefaoui M, Bouraiou A, et al. Photovoltaic output power performance assessment and forecasting: Impact of meteorological variables. Solar Energy. 2021;**220**:745-757. DOI: 10.1016/j.solener.2021.04.004

[7] Chahboun S, Maaroufi M. Principal component analysis and machine learning approaches for photovoltaic power prediction: A comparative study. Applied Sciences. 2021;**11**(17):7943. DOI: 10.3390/app11177943

[8] Malvoni M, De Giorgi MG, Congedo PM. Photovoltaic forecast based on hybrid PCA–LSSVM using dimensionality reducted data. Neurocomputing. 2016;**211**:72-83. DOI: 10.1016/j.neucom.2016.01.104

[9] Malvoni M, De Giorgi MG, Congedo PM. Forecasting of PV power generation using weather input data-preprocessing techniques. Energy Procedia. 2017;**126**:651-658. DOI: 10.1016/j.egypro.2017.08.293

[10] Atkinson AC, Riani M, Corbellini A. The box–cox transformation: Review and extensions. Statistical Science. 2021; **36**(2):239-255

[11] Uribe DR. Short-Term Solar Power Forecasting Using Different Machine Learning Models. 2020

[12] R Core Team. R: A language and environment for statistical computing. R Foundation for Statistical Computing, Vienna, Austria. 2018. Available online at https://www.R-project.org/

[13] Fraccanabbia N, Da Silva RG, Ribeiro MHDM, Moreno SR, Dos Santos Coelho L, Mariani VC. Solar power forecasting based on ensemble learning methods. In: Proc Int Jt Conf Neural Networks. 2020

[14] Zhou J, Li E, Wei H, Li C, Qiao Q, Armaghani DJ. Random forests and cubist algorithms for predicting shear strengths of rockfill materials. Applied Sciences. 2019;**9**(8):1-16

[15] Wuttichaikitcharoen P, Babel MS. Principal component and multiple regression analyses for the estimation of suspended sediment yield in ungauged basins of northern Thailand. Watermark. 2014;**6**(8):2412-2435

Chapter 5

Variable Selection in Nonlinear Principal Component Analysis

Hiroko Katayama, Yuichi Mori and Masahiro Kuroda

Abstract

Principal components analysis (PCA) is a popular dimension reduction method and is applied to analyze quantitative data. For PCA to qualitative data, nonlinear PCA can be applied, where the data are quantified by using optimal scaling that nonlinearly transforms qualitative data into quantitative data. Then nonlinear PCA reveals nonlinear relationships among variables with different measurement levels. Using this quantification, we can consider variable selection in the context of PCA for qualitative data. In PCA for quantitative data, modified PCA (M.PCA) of Tanaka and Mori derives principal components which are computed as a linear combination of a subset of variables but can reproduce all the variables very well. This means that M.PCA can select a reasonable subset of variables with different measurement levels if it is extended so as to deal with qualitative data by using the idea of nonlinear PCA. A nonlinear M.PCA is therefore proposed for variable selection in nonlinear PCA. The method, in this chapter, is based on the idea in "Nonlinear Principal Component Analysis and its Applications" by Mori et al. (Springer). The performance of the method is evaluated in a numerical example.

Keywords: quantification, categorical data, modified PCA, stepwise selection, cumulative proportion, RV-coefficient

1. Introduction

Principal components analysis (PCA) is a popular dimension reduction method and is applied to analyze quantitative data. For PCA to qualitative data, the data are quantified by using optimal scaling that nonlinearly transforms qualitative data into quantitative data. The PCA with optimal scaling is called nonlinear PCA. Nonlinear PCA reveals all qualitative variables uniformly as numerical variables by using optimal scaling quantifiers in the analysis, that is, it can deal with nonlinear relationships among variables with different measurement levels.

Using this quantification, we can consider variable selection in the context of PCA for qualitative data. In PCA for quantitative data, Tanaka and Mori discussed a method called modified PCA (M.PCA) that can be used to compute principal components (PCs) using only a selected subset of variables that represents all of the variables, including those not selected [1]. Since M.PCA includes variable selection procedures in the analysis, if we quantify all the qualitative variables by using the optimal scaling and then apply M.PCA to the quantified data, we can select a reasonable subset of variables from the qualitative data.

In this chapter, we refer to Mori et al. [2] to revisit a variable selection problem in PCA for qualitative data. The proposed method here (we call it nonlinear M.PCA or NL.M.PCA) is an extension of M.PCA so as to deal with a mixture of quantitative and qualitative data. In Section 2 we provide the overview of NL.M.PCA (optimization, the original M.PCA and NL.M.PCA for qualitative data) based on studies by Mori et al. [2], and in Section 3, we apply this method to the customer engagement data [3] to show how it works in the real data and how you use the output from the method for variable selection, and to evaluate the performance of the method.

2. Modified PCA for mixed measurement level data

2.1 Quantification of qualitative data

We must use a suitable quantification method in the context of PCA because we here wish to consider a variable selection problem in PCA. One of the best methods is the optimal scaling in nonlinear PCA. Nonlinear PCA is a method to deal with qualitative data, which estimates the parameters of PCA and quantifies qualitative variables simultaneously by alternating between estimation and quantification. PRINCIPALS of Young et al. [4] and PRINCIPALS of Gifi [5] are algorithms for nonlinear PCA. Here we use PRINCIPALS.

PRINCIPALS is an algorithm using the alternating least squares (ALS) algorithm as follows: Let $\mathbf{Y} = (\mathbf{y}_1 \mathbf{y}_2 \dots \mathbf{y}_p)$ be a data matrix of n objects by p categorical variables and let $\mathbf{y}_j$ of $\mathbf{Y}$ be a qualitative vector with K_j categories labeled $1, \dots, K_j$. PRINCIPALS minimizes the loss function

$$\sigma_L(\mathbf{Z}, \mathbf{A}, \mathbf{Y}^*) = \mathrm{tr}\left(\mathbf{Y}^* - \hat{\mathbf{Y}}\right)^\top\left(\mathbf{Y}^* - \hat{\mathbf{Y}}\right) = \mathrm{tr}\left(\mathbf{Y}^* - \mathbf{Z}\mathbf{A}^\top\right)^\top\left(\mathbf{Y}^* - \mathbf{Z}\mathbf{A}^\top\right), \quad (1)$$

where $\mathbf{Y}^*$ is an optimally scaled matrix form $\mathbf{Y}$, $\mathbf{Z}$ is an $n \times r$ matrix of n component scores on r $(1 \le r \le p)$ components, and $\mathbf{A} = \{\mathbf{a}_1 \mathbf{a}_2 \dots \mathbf{a}_r\}$ is a $p \times r$ weight matrix that gives the coefficients of the linear combinations. PRINCIPALS alternately makes two estimations: the model parameters $\mathbf{Z}$ and $\mathbf{A}$ for ordinary PCA, and the data parameter for optimally scaled data $\mathbf{Y}^*$.

In the computation of PRINCIPALS, $\mathbf{Y}^*$ are standardized for each variable such as to satisfy restrictions $\mathbf{Y}^{*\top}\mathbf{1}_n = \mathbf{0}_p$ and $\mathrm{diag}\left[\frac{\mathbf{Y}^{*\top}\mathbf{Y}^*}{n}\right] = \mathbf{I}_p$. We denote the value θ estimated the t-th iteration by $\theta^{(t)}$. Given the initial data $\mathbf{Y}^{*(0)}$ (the observed data $\mathbf{Y}$ may be used as $\mathbf{Y}^{*(0)}$ after the above standardization), PRINCIPALS iterates the following two steps:

- *Model estimation step*: By solving the eigenvalue problem (EVP) of the covariance matrix of $\mathbf{Y}^{*(t)}$ $(= \mathbf{S})$

$$[\mathbf{S} - \lambda\mathbf{I}]\mathbf{a} = 0, \quad (2)$$

where λ is the eigenvalues, obtain $\mathbf{A}^{(t+1)}$ and compute $\mathbf{Z}^{(t+1)} = \mathbf{Y}^{*(t)}\mathbf{A}^{(t+1)}$. Update $\hat{\mathbf{Y}}^{(t+1)} = \mathbf{Z}^{(t+1)}\mathbf{A}^{(t+1)\top}$.

- *Optimal scaling step*: Obtain $\mathbf{Y}^{*(t+1)}$ such that

$$\mathbf{Y}^{*(t+1)} = \arg\min_{\mathbf{Y}^{*(t)}} \mathrm{tr}\left(\mathbf{Y}^{*(t)} - \hat{\mathbf{Y}}^{(t+1)}\right)^{\top}\left(\mathbf{Y}^{*(t)} - \hat{\mathbf{Y}}^{(t+1)}\right) \tag{3}$$

for fixed $\hat{\mathbf{Y}}^{(t+1)}$ by separately estimating $\mathbf{y}_j^*$ for each variable j under the measurement restrictions on each of the variables. That is, compute $\mathbf{q}_j^{(t+1)}$ for nominal variables as

$$\mathbf{q}_j^{(t+1)} = \left(\mathbf{G}_j^{\top}\mathbf{G}_j\right)^{-1}\mathbf{G}_j^{\top}\hat{\mathbf{y}}_j^{(t+1)}, \tag{4}$$

where $\mathbf{q}_j$ is a $K_j \times 1$ category score vector for $\mathbf{y}_j^*$ and $\mathbf{G}_j$ is an $n \times K_j$ indicator matrix

$$\mathbf{G}_j = \left(g_{jik}\right) = \begin{pmatrix} g_{j11} & \cdots & g_{j1K_j} \\ \vdots & \vdots & \vdots \\ g_{jn1} & \cdots & g_{jnK_j} \end{pmatrix} = \left(\mathbf{g}_{j1} \cdots \mathbf{g}_{jK_j}\right), \tag{5}$$

where

$$g_{jik} = \begin{cases} 1 & \text{if object } i \text{ belongs to category } k \\ 0 & \text{if object } i \text{ belongs to some other category } k'(\neq k), \end{cases} \tag{6}$$

and then the optimally scaled vector $\mathbf{y}_j^*$ is obtained by $\mathbf{y}_j^* = \mathbf{G}_j\mathbf{q}_j$.

Re-compute $\mathbf{q}_j^{(t+1)}$ for ordinal variables using the monotone regression [6]. For nominal and ordinal variables, update $\mathbf{y}_j^{*(t+1)} = \mathbf{G}_j\mathbf{q}_j^{(t+1)}$ and standardize $\mathbf{y}_j^{*(t+1)}$. For numerical variables, standardize the observed vector $\mathbf{y}_j$ and set $\mathbf{y}_j^{*(t+1)} = \mathbf{y}_j$.

These two steps alternately iterate until convergence, and $\mathbf{y}_j^*$ obtained at convergence is the quantified variable while $\mathbf{A}$ and $\mathbf{Z}$ are the solutions of PCA for qualitative data.

2.2 Modified PCA

M.PCA of Tanaka and Mori [1] derives PCs that are computed using only a selected subset but represent all of the variables, including those not selected. This means that M.PCA naturally includes variable selection procedures in its estimation process. Although there are several variable selection methods in PCA, we use M.PCA, because a subset of variables selected by M.PCA can represent all the variables very well and it is easy to incorporate the quantification method in Section 2.1 into M.PCA, which will be described in Section 2.3.

Suppose we obtain an $n \times p$ data matrix $\mathbf{Y}$ that consists of numerical variables or optimally quantified variables. Let $\mathbf{Y}$ be decomposed into an $n \times q$ submatrix $\mathbf{Y}_1$ and an $n \times (p - q)$ submatrix $\mathbf{Y}_2$ $(1 \le q \le p)$. $\mathbf{Y}$ is represented by r PCs, which is a linear combination of a submatrix $\mathbf{Y}_1$, that is, $\mathbf{Z} = \mathbf{Y}_1\mathbf{A}$, where r is the number of PCs $(1 \le r \le q)$. To derive $\mathbf{A} = (\mathbf{a}_1 \ldots \mathbf{a}_r)$, the following Criterion 1 based on Rao [7] and Criterion 2 based on Robert and Escoufier [8] can be used:

(Criterion 1) The prediction efficiency $\mathbf{Y}$ is maximized using a linear predictor in terms of $\mathbf{Z}$.

(Criterion 2) The closeness of configurations between $\mathbf{Y}$ and $\mathbf{Z}$ is maximized using the RV-coefficient.

We denote the covariance matrix of $\mathbf{Y} = (\mathbf{Y}_1, \mathbf{Y}_2)$ as $\mathbf{S} = \begin{pmatrix} \mathbf{S}_{11} & \mathbf{S}_{12} \\ \mathbf{S}_{21} & \mathbf{S}_{22} \end{pmatrix}$, where the subscript i of $\mathbf{S}$ corresponds to $\mathbf{Y}_i$. The maximization criteria for the above Criterion 1 and Criterion 2 are given by the proportion P

$$P = \sum_{j=1}^{r} \lambda_j / \mathrm{tr}\,(\mathbf{S}), \tag{7}$$

and the RV-coefficient

$$RV = \left\{ \sum_{j=1}^{r} \lambda_j^2 / \mathrm{tr}\left(\mathbf{S}^2\right) \right\}^{1/2}, \tag{8}$$

respectively, where λ_j is the j-th eigenvalue with the order of magnitude of the EVP

$$\left[\left(\mathbf{S}_{11}^2 + \mathbf{S}_{12}\mathbf{S}_{21}\right) - \lambda \mathbf{S}_{11}\right]\mathbf{a} = 0. \tag{9}$$

The solution is obtained as a matrix $\mathbf{A}$, the columns of which consist of the eigenvectors associated with the largest r eigenvalues of EVP (9), and $\mathbf{Y}_1$ that provides the largest value of P or RV is the best subset of q variables among all possible subsets of size q. Thus, to obtain a reasonable subset of variables with size q in PCA, you apply M.PCA to the data and find the subset of size q, $\mathbf{Y}_1$, that has the largest P or RV. The selected subset $\mathbf{Y}_1$ is reasonable in the sense of PCA because it contains information that includes not only the selected variables $\mathbf{Y}_1$ but also the deleted ones $\mathbf{Y}_2$.

2.3 Modified PCA for mixed measurement level data

M.PCA is a good method to find a reasonable subset of numerical variables as described in the previous section. To select variables from mixed measurement level data by using a criterion in M.PCA, qualitative/categorical variables in the data should be quantified in an appropriate manner. Based on the original idea in ref. [9], considering PRINCIPALS in Section 2.1 and M.PCA in Section 2.2, it is easy to incorporate the quantification (PRINCIPALS) into M.PCA, because we can formulate M.PCA for qualitative data only by replacing the EVP (2)) in the *Model estimation step* of PRINCIPALS by the EVP (9) to get the model parameters $\mathbf{A}$ and $\mathbf{Z}$ for M.PCA. Thus, M.PCA and optimal scaling are alternately executed until $\theta^* = \mathrm{tr}\left(\mathbf{Y}^* - \hat{\mathbf{Y}}\right)^\top\left(\mathbf{Y}^* - \hat{\mathbf{Y}}\right) = \mathrm{tr}\left(\mathbf{Y}^* - \mathbf{Z}\mathbf{A}^\top\right)^\top\left(\mathbf{Y}^* - \mathbf{Z}\mathbf{A}^\top\right)$ is minimized. This is nonlinear M.PCA or NL.M.PCA.

Here, we rewrite the ALS algorithm of PRINCIPALS as follows—for given initial data $\mathbf{Y}^{*(0)} = \left(\mathbf{Y}_1^{*(0)},\ \mathbf{Y}_2^{*(0)}\right)$ from the original data $\mathbf{Y}$, the following two steps are iterated until convergence:

- *Model estimation step*: From $\mathbf{Y}^{*(t)} = \left(\mathbf{Y}_1^{*(t)},\ \mathbf{Y}_2^{*(t)}\right)$, obtained $\mathbf{A}^{(t)}$ by solving the EVP (9).

  Compute $\mathbf{Z}^{(t)}$ from $\mathbf{Z}^{(t)} = \mathbf{Y}_1^{*(t)}\mathbf{A}^{(t)}$. Update $\hat{\mathbf{Y}}^{(t+1)} = \mathbf{Z}^{(t)}\mathbf{A}^{(t)}$.

- *Optimal scaling step*: Obtain $\mathbf{Y}^{*(t+1)}$ for fixed $\hat{\mathbf{Y}}^{(t+1)}$ by separately estimating $\mathbf{y}_j^*$ ($=\mathbf{G}_j\mathbf{q}_j$) for each variable j under the measurement restrictions. Re-compute $\mathbf{Y}_j^{*(t+1)}$ by an additional transformation to keep the monotonicity restriction for ordinal variables and skip this computation for numerical variables.

$\mathbf{Y}^* = (\mathbf{Y}_1^*,\ \mathbf{Y}_2^*)$ obtained after convergence is an optimally scaled (quantified) matrix of $\mathbf{Y}$, and $\mathbf{Y}_1$ corresponding to $\mathbf{Y}_1^*$ is a subset to be selected and $\mathbf{Y}_2$ to $\mathbf{Y}_2^*$ is one to be deleted.

NL.M.PCA procedure for fixed q is as described above, but since the variable selection performs M.PCA calculation for $q = p, \ldots, r$ and ${}_pC_q$ times to find the best $\mathbf{Y}_1$, there are three possible *type*s of selection according to where the quantification is implemented in the computation flow (see Fig. 4.1 in [2]).

The first type (*Type 1*) is that the quantification is performed only once at first, that is, nonlinear PCA is applied to the data $\mathbf{Y}$ to obtain the quantified data $\mathbf{Y}^*$, and ordinary M.PCA selection is applied to $\mathbf{Y}^*$. No more quantification is carried out in the selection stage. The second type (*Type 2*) is that the quantification is carried out every time after the best subset of size q is found in the selection stage. That is, the quantified $(\mathbf{Y}_1^*, \mathbf{Y}_2^*)$ based on the best subset of the size q found in the previous selection is used to find the best subset of size $q-1$ or $q+1$ in the next selection. The third type (*Type 3*) is that the quantification is carried out for every temporary $(\mathbf{Y}_1,\ \mathbf{Y}_2)$ in the section stage, that is, NL.M.PCA is performed whenever temporary $(\mathbf{Y}_1,\ \mathbf{Y}_2)$ is given to compute its criterion value.

A reasonable subset of size q is given as $\mathbf{Y}_1$ corresponding to the best subset $\mathbf{Y}_1^*$ which is finally found at q when the selection procedure is terminated.

3. A numerical example

3.1 Data

The data we analyze here was gathered in the survey about the relationship among customer engagement on "fashion," "brand," and "shop staff" [3]. The questions (variables) are divided into three groups based on the purposes for consumption: "Involvement" (16 variables), "Expectations" (35 variables), and "Values" (34 variables). The total number of questions is 85 on a five level scale and 825 responses are obtained. Ohyabu et al. [3] analyzed this data to find the structure of the customer consciousness, but we use this data simply as sample data for variable selection in PCA without considering the original purpose in ref. [3]. Here we apply NL.M.PCA to the second question group "Expectation" (35 variables) to show the performance of the proposed method. The questions asked in the survey are indicated in the "Question" column of **Table 1** and answers (responses) are shown in **Table 2**.

Group	Item	Question	q = 25
About the fashion	Q	We would like to ask you about your thoughts and behaviors about fashion.	
	Q1	I think about fashion by putting on clothes or choosing the clothes.	×
	Q2	I think about fashion by putting on clothes or choosing the clothes.	×
	Q3	I want to know about fashion by putting on clothes or choosing clothes.	×
	Q4	I'm enthusiastic when I think about fashion.	×
	Q5	I'm happy with thinking about fashion.	×
	Q6	I feel relaxed when I think about fashion.	×
	Q7	I'm proud of my fashions when I think about fashion.	×
	Q8	I spend a lot of one's time when I think about the fashions.	
	Q9	I talk about fashion with my friends.	
	Q10	I'm checking about SNS or writing comment for fashion.	×
	Q11	I'm posting about a fashion to SNS.	×
About the brand	Q	We would like to ask you about your thoughts and behaviors about fashion brands.	
	Q12	I think about the brand by putting on clothes or choosing the clothes.	
	Q13	I think about the brand by putting on clothes or choosing the clothes.	×
	Q14	I want to know about the brand by putting on clothes or choosing the clothes.	×
	Q15	I'm enthusiastic when I think about the brand.	×
	Q16	I'm happy when I think about the brand.	
	Q17	I feel relaxed good when I think about the brand.	×
	Q18	I'm proud when I think about the brand.	×
	Q19	I spend a lot of one's time when I think about the brand.	×
	Q20	I always use a specific brand when I wear or choose clothes.	
	Q21	I always use the brand when I clothes or choice of clothes.	×
	Q22	I'm checking about SNS or writing comment for the brand.	×
	Q23	I'm posting about a brand to an SNS of mine.	×
About the shop staff	Q	We would like to ask you about your thoughts and behaviors about the staff member	
	Q24	I think about the staff member by talking to other staff	×
	Q25	I think about staff members when I speak to other shop staff.	
	Q26	I want to know more about shop staff by speaking.	×
	Q27	I'm enthusiastic when I'm talking with staff members.	
	Q28	I'm happy when I'm talking with staff members.	×
	Q29	I feel relaxed when I'm talking with staff members.	×
	Q30	I'm proud when I'm talking with shop staff.	
	Q31	I spend a lot of time talking with shop staff.	×
	Q32	I always talk to the specific staff member when choosing clothes or putting on clothes.	

Group	Item	Question	q = 25
	Q33	I always talk the specific staff member.	×
	Q34	I'm checking about the specific staff member of SNS or writing comment for the brand.	
	Q35	I'm posting about the specific staff member to my SNS.	×

Table 1.
35 questions in "expectation" and 25 selected ones (marked by × in the right column).

Respondent	Q1	Q2	Q3	Q4	Q5	Q6	Q7	Q8	Q9	Q10	Q11	Q12	Q13	Q14	Q15	Q16	Q17	Q18	Q19	Q20	Q21	Q22	Q23	Q24	Q25	Q26	Q27	Q28	Q29	Q30	Q31	Q32	Q33	Q34	Q35
1	1	1	1	1	1	1	3	1	1	1	1	2	2	2	2	2	2	3	2	2	3	3	4	2	2	3	3	3	2	4	2	3	2	5	5
2	1	1	1	1	1	1	1	1	3	3	3	1	1	1	1	1	1	1	1	1	2	4	3	1	1	2	1	1	1	1	1	1	1	3	3
3	2	2	2	2	2	2	4	3	2	3	4	2	2	2	2	2	2	4	3	3	3	4	4	5	5	5	5	5	3	5	5	5	5	5	5
4	1	1	2	1	1	2	1	1	1	2	3	1	2	2	2	2	1	2	1	1	1	3	4	1	1	1	1	1	1	1	2	2	2	2	3
5	2	2	3	2	2	3	3	3	3	3	5	3	3	3	3	3	3	3	3	5	3	5	3	5	5	5	5	5	5	5	5	5	5	5	4
6	4	4	4	4	4	5	4	4	5	5	5	4	5	4	4	4	4	5	4	4	4	4	4	4	4	4	5	4	4	4	5	4	4	4	3
7	2	3	1	1	1	1	2	1	2	1	1	2	2	2	2	2	1	1	1	1	2	1	2	2	3	3	3	1	3	2	1	1	1	4	1
8	1	1	1	1	1	1	1	1	1	1	1	1	1	1	1	2	1	2	1	1	1	2	1	5	5	5	5	5	5	5	5	5	5	5	5
9	2	2	2	3	2	2	3	4	3	5	5	2	2	3	3	3	3	3	2	2	2	4	5	4	4	4	5	3	3	4	4	3	3	5	5
10	1	1	1	1	1	1	2	1	4	3	5	2	3	1	1	1	1	3	1	1	1	3	5	4	4	4	3	2	4	5	5	4	4	4	5
11	2	2	2	3	3	3	3	3	3	3	3	3	3	2	3	3	3	3	3	3	3	3	3	3	3	3	3	3	3	3	3	5	5	5	5
12	1	1	2	1	1	1	4	3	2	4	4	2	2	2	1	1	1	3	2	2	1	4	4	2	4	4	2	2	2	4	3	3	2	4	4
13	2	2	2	2	1	2	2	3	3	3	3	2	2	2	2	2	2	2	2	2	2	2	3	2	1	2	1	2	2	2	1	2	2	2	4
14	1	1	1	2	2	2	4	2	1	1	4	2	2	1	1	1	1	1	1	1	1	1	2	2	2	1	1	1	1	1	1	1	5	2	2
15	2	2	3	2	1	2	4	3	3	2	5	2	2	2	2	2	1	2	2	2	2	3	3	2	1	4	2	2	2	2	1	2	2	5	5
16	1	1	2	2	2	2	3	3	2	2	4	2	2	3	2	2	1	2	2	2	1	4	4	5	5	5	2	3	3	3	3	5	2	5	5
17	1	1	1	1	1	1	2	2	1	2	3	2	2	1	1	1	1	1	2	3	2	3	4	4	3	4	4	4	4	4	4	3	4	4	5
18	2	2	3	2	2	2	4	4	3	5	5	2	2	3	2	2	2	2	3	3	2	5	5	5	5	1	5	5	5	5	5	5	5	5	5
19	2	2	2	2	2	2	4	3	3	3	3	3	3	2	2	2	2	3	3	3	3	3	3	4	4	4	4	3	3	4	4	4	3	4	4
20	2	2	2	2	2	2	2	2	2	2	4	2	2	2	2	2	2	2	2	2	2	2	4	2	2	2	2	2	2	2	2	2	2	4	5
21	1	1	1	1	1	1	3	2	1	1	5	1	1	1	1	1	1	1	1	1	1	1	5	1	1	1	1	1	1	1	1	1	1	5	5
22	1	1	1	1	1	1	3	2	3	2	3	1	1	1	1	1	1	1	2	3	3	2	3	3	3	4	3	3	3	4	4	4	4	4	5
23	2	1	2	1	1	1	1	3	4	2	5	2	1	3	1	1	1	1	2	3	1	2	5	3	4	5	2	2	2	4	3	3	2	5	5
24	1	1	1	1	1	1	1	1	1	1	1	1	1	1	1	1	2	1	1	1	1	1	1	1	1	1	1	1	1	1	1	2	1	1	1
25	1	1	1	1	2	1	2	1	1	1	1	1	1	1	1	1	2	1	1	1	1	1	1	1	1	1	1	1	1	1	1	1	1	1	1
26	1	1	1	2	2	1	3	3	2	2	3	1	2	2	1	1	1	2	3	2	1	2	4	2	3	3	2	3	3	4	3	2	2	5	5
27	1	1	1	2	1	1	2	2	2	2	3	1	1	1	2	1	1	2	1	2	1	2	3	1	1	1	1	1	1	2	3	2	3	2	3
28	2	2	1	3	3	3	3	2	1	4	3	2	2	1	2	2	2	1	2	2	1	3	4	2	2	1	1	1	1	2	1	2	1	4	2
29	1	1	1	1	1	1	1	1	1	2	2	2	2	1	2	1	1	1	2	1	1	1	2	2	1	2	2	1	1	2	2	2	2	3	4
30	2	2	2	2	2	2	2	2	2	3	5	3	3	3	2	3	2	3	2	3	3	2	2	3	3	3	3	3	3	3	3	3	2	5	5
:	:	:	:	:	:	:	:	:	:	:	:	:	:	:	:	:	:	:	:	:	:	:	:	:	:	:	:	:	:	:	:	:	:	:	:
:	:	:	:	:	:	:	:	:	:	:	:	:	:	:	:	:	:	:	:	:	:	:	:	:	:	:	:	:	:	:	:	:	:	:	:
824	3	1	1	1	2	4	3	1	2	4	4	1	2	2	5	3	3	1	1	2	1	1	3	2	1	1	4	3	2	4	1	1	1	1	1
825	3	3	1	1	1	5	2	1	4	3	1	1	5	1	4	2	2	2	2	2	2	2	2	3	2	2	2	3	3	3	3	2	2	3	3

Table 2.
Expectation data (825 responses on 35 variables).

3.2 Output from NL.M.PCA

Table 3 shows the output of NL.M.PCA when NL.M.PCA is applied to Expectation data with $r = 5$ of the number of PCs, proportion P as a criterion, and forward-backward stepwise selection and *type* 3 quantifications as selection procedures. The number q is the number of selected variables and the value P is the criterion value. $\mathbf{Y}_1|\mathbf{Y}_2$ shows that the left side of each row is the question numbers to be selected ($\mathbf{Y}_1$) and the right side to be deleted ($\mathbf{Y}_2$). If you have a specific number q for variables to be used, such as 20, 10, or 2/3 = 24, 1/2 = 18, you can use variables whose numbers are displayed in $\mathbf{Y}_1$ at that q. If the number of variables to be used is not determined, the proportion P can be used. For example, since the proportion P is 66.95% with all 35 variables, if you want to keep P up to 65%, looking at the row of $P = 0.6512$ (i.e., $q = 20$), you can use 20 variables in $\mathbf{Y}_1$. Alternatively, if the difference between the proportion with all 35 variables and that with selected variables should be less than 1%, 25 variables can be used because $0.6695 - 0.01 = 0.6595$, which is the P value at

q																	Y1	Y2																		*P*
35	1	2	3	4	5	6	7	8	9	10	11	12	13	14	15	16	17	18	19	20	21	22	23	24	25	26	27	28	29	30	31	32	33	34	35	0.6695
34	1	2	3	4	5	6	7	8	9	10	11	12	13	14	15	16	17	18	19	20	21	22	23	24	26	27	28	29	30	31	32	33	34	35│	25	0.6689
33	1	2	3	4	5	6	7	8	9	10	11	12	13	14	15	16	17	18	19	20	21	22	23	24	26	27	28	29	30	31	33	34	35│	25	32	0.6682
32	1	2	3	4	5	6	7	8	9	10	11	12	13	14	15	16	17	18	19	20	21	22	23	24	26	28	29	30	31	33	34	35│	25	27	32	0.6675
31	1	2	3	4	5	6	7	9	10	11	12	13	14	15	16	17	18	19	20	21	22	23	24	26	28	29	30	31	33	34	35│	8	25	27	32	0.6666
30	1	2	3	4	5	6	7	9	10	11	12	13	14	15	17	18	19	20	21	22	23	24	26	28	29	30	31	33	34	35│	8	16	25	27	32	0.6656
29	1	2	3	4	5	6	7	9	10	11	12	13	14	15	17	18	19	20	21	22	23	24	26	28	29	31	33	34	35│	8	16	25	27	30	32	0.6646
28	1	2	3	4	5	6	7	9	10	11	13	14	15	17	18	19	20	21	22	23	24	26	28	29	31	33	34	35│	8	12	16	25	27	30	32	0.6636
27	1	2	3	4	5	6	7	9	10	11	13	14	15	17	18	19	20	21	22	23	24	26	28	29	31	33	35│	8	12	16	25	27	30	32	34	0.6624
26	1	2	3	4	5	6	7	10	11	13	14	15	17	18	19	20	21	22	23	24	26	28	29	31	33	35│	8	9	12	16	25	27	30	32	34	0.6611
25	1	2	3	4	5	6	7	10	11	13	14	15	17	18	19	21	22	23	24	26	28	29	31	33	35│	8	9	12	16	20	25	27	30	32	34	0.6598
24	1	2	3	4	6	7	10	11	13	14	15	17	18	19	21	22	23	24	26	28	29	31	33	35│	5	8	9	12	16	20	25	27	30	32	34	0.6583
23	1	2	3	4	6	7	10	11	13	14	15	17	18	19	21	22	23	24	26	29	31	33	35│	5	8	9	12	16	20	25	27	28	30	32	34	0.6568
22	1	2	3	4	6	7	10	11	13	14	15	17	18	19	21	22	24	26	29	31	33	35│	5	8	9	12	16	20	23	25	27	28	30	32	34	0.6552
21	2	3	4	6	7	10	11	13	14	15	17	18	19	21	22	24	26	29	31	33	35│	1	5	8	9	12	16	20	23	25	27	28	30	32	34	0.6533
20	2	3	4	6	7	10	11	13	14	17	18	19	21	22	24	26	29	31	33	35│	1	5	8	9	12	15	16	20	23	25	27	28	30	32	34	0.6512
19	2	3	4	6	7	10	11	13	14	17	18	19	21	22	26	29	31	33	35│	1	5	8	9	12	15	16	20	23	24	25	27	28	30	32	34	0.6492
18	2	3	4	6	7	10	11	13	14	17	19	21	22	26	29	31	33	35│	1	5	8	9	12	15	16	18	20	23	24	25	27	28	30	32	34	0.6467
17	2	4	6	7	10	11	13	14	17	19	21	22	26	29	31	33	35│	1	3	5	8	9	12	15	16	18	20	23	24	25	27	28	30	32	34	0.6435
16	1	4	6	7	10	11	13	14	15	19	21	22	25	27	32	35│	2	3	5	8	9	12	16	17	18	20	23	24	26	28	29	30	31	33	34	0.6403
15	1	4	6	7	10	11	14	15	19	21	22	25	27	32	35│	2	3	5	8	9	12	13	16	17	18	20	23	24	26	28	29	30	31	33	34	0.6370
14	1	6	7	10	11	14	15	19	21	22	25	27	32	35│	2	3	4	5	8	9	12	13	16	17	18	20	23	24	26	28	29	30	31	33	34	0.6326
13	1	6	7	10	11	14	15	19	21	25	27	32	35│	2	3	4	5	8	9	12	13	16	17	18	20	22	23	24	26	28	29	30	31	33	34	0.6278
12	1	6	10	11	14	15	19	21	25	27	32	35│	2	3	4	5	7	8	9	12	13	16	17	18	20	22	23	24	26	28	29	30	31	33	34	0.6224
11	1	6	10	11	14	15	19	25	27	32	35│	2	3	4	5	7	8	9	12	13	16	17	18	20	21	22	23	24	26	28	29	30	31	33	34	0.6154
10	1	6	10	11	14	19	25	27	32	35│	2	3	4	5	7	8	9	12	13	15	16	17	18	20	21	22	23	24	26	28	29	30	31	33	34	0.6068
9	1	6	10	11	14	19	25	27	35│	2	3	4	5	7	8	9	12	13	15	16	17	18	20	21	22	23	24	26	28	29	30	31	32	33	34	0.5968
8	1	6	11	14	19	25	27	35│	2	3	4	5	7	8	9	10	12	13	15	16	17	18	20	21	22	23	24	26	28	29	30	31	32	33	34	0.5870
7	1	6	11	14	19	25	32│	2	3	4	5	7	8	9	10	12	13	15	16	17	18	20	21	22	23	24	26	27	28	29	30	31	33	34	35	0.5714
6	1	6	11	14	25	32│	2	3	4	5	7	8	9	10	12	13	15	16	17	18	19	20	21	22	23	24	26	27	28	29	30	31	33	34	35	0.5400
5	6	11	14	21	27│	1	2	3	4	5	7	8	9	10	12	13	15	16	17	18	19	20	22	23	24	25	26	28	29	30	31	32	33	34	35	0.5319

Table 3.
Selection results (expectations, $r = 5$, proportion P, forward-backward stepwise selection, Type 3).

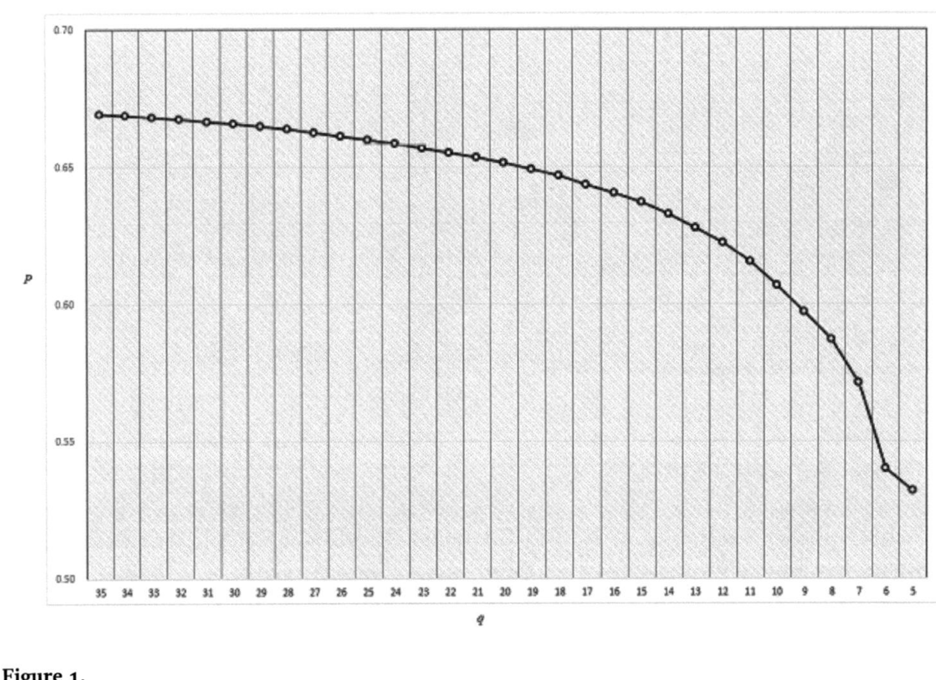

Figure 1.
Change of the proportion P for every q (from 35 to 5).

$q = 25$. **Figure 1** shows the change of P for every q. This graph can be used to obtain guidance on the determination of the number of variables. Looking at this graph, if there is a large drop in P, the number of variables just before that point can be used (for this data, no particular drop is observed).

When using RV, the same considerations are applied, and scatter plots are also considered to see how close the configurations are.

3.3 Results of variable selection

Here we select a subset of variables from 35 variables of Expectation data focusing on the loss of proportion P. Suppose we want to keep it under 1%, $q = 25$ which is assigned from **Table 3** and **Figure 1**. The selected variables are marked by $\times$ in the right column of **Table 1**. As far as looking at the variables deleted from each block, two variables {8, 9} from 11 variables in "fashion" block, three variables {12, 16, 20} from 12 variables in "brand" block, and five variables {25, 27, 30, 32, 34} from 12 variables in "shop staff" block are deleted. That is, nine variables are selected from the first two blocks and seven from the third block. It can be stated that the proposed method selects a reasonable subset of variables. Comparing the number of deleted variables in the three blocks, a slightly larger number of variables are removed from the third block, so it is thought that questions on "shop staff" have little information rather than those in the other two blocks and some of them have less significance on the prediction efficiency. From this point of view, we can evaluate the usefulness of each question in the questionnaire.

To evaluate the significance of variables, we observe how many times each variable is selected through the selection for $q = 35, \ldots, 5$. Extracting the variables selected over 2/3 times (24 or more), for example, in the "fashion" block, variables {1, 6, 10, 11} were selected. Given the fact that the close-up questions are located close to each other (1 to 3, recognition on fashion, 4 to 7—consciousness on fashion, 8 to 11—activity on fashion), it is generally clear that NL.M.PCA using the proportion P selects variables well-balanced from the close-up questions. Similarly, if the most frequently selected variables (such as the above four items) are considered as the most important questions, they should be involved in future surveys. If variables are selected a few times, they should not be involved in the future. In such a way, there is a possibility to use the selection results to evaluate the questionnaire itself.

4. Concluding remarks

We reconsider a variable selection problem in PCA for qualitative data based on the idea of Mori et al. [2]. For the problem of how to deal with qualitative data, we apply optimal scaling with the ALS algorithm [4] to the qualitative data. For the variable selection in PCA, we use the criteria in M.PCA of Tanaka and Mori [1] for optimally quantified data. That is, the proposed method is an extension of M.PCA by implementing optimal scaling into M.PCA so as to select a subset of qualitative variables. Using this method, since the quantification is done separately for each variable, we can select a subset of variables from mixed measurement level data.

We apply this method to real data from a customer engagement study [3] to select a subset of qualitative variables by using a criterion that maximizes the prediction efficiency. For a case where there is no preassigned number of variables to be selected, it can be suggested to specify the number in such a way that the maximum loss of the efficiency is not over a certain percentage.

As a result, variables are selected in a well-balanced manner from questions asking similar contents, and the selected subset, therefore, provides as much information as possible. It is expected that the nonlinear M.PCA works well for any mixed measurement level data.

Author details

Hiroko Katayama[*†], Yuichi Mori[2†] and Masahiro Kuroda[2]

1 Graduate School of Informatics, Okayama University of Science, Okayama, Japan

2 Department of Management, Okayama University of Science, Okayama, Japan

*Address all correspondence to: hk22@pub.ous.ac.jp

† These authors contributed equally.

References

[1] Tanaka Y, Mori Y. Principal component analysis based on a subset of variables: Variable selection and sensitivity analysis. American Journal of Mathematics and Management Sciences. 1997;17(1 & 2):61-89

[2] Mori Y, Kuroda M, Makino N. Nonlinear Principal Component Analysis and its Applications (JSS Research Series in Statistics). Singapore: Springer; 2017

[3] Ohyabu R, Kuroda M, Seino S, Zhang Z. Exploring interplay among customer engagements with multiple objects, the 6th Naples forum on service. Service Dominant Logic, Network & Systems Theory and Service Science: Integrating three Perspectives for a New Service Agenda. 2019:103

[4] Young FW, Takane Y, de Leeuw J. Principal components of mixed measurement level multivariate data: An alternating least squares method with optimal scaling features. Psychometrika. 1978;**43**:279-281

[5] Gifi A. Nonlinear Multivariate Analysis. Chichester: Wiley; 1990

[6] Kruskal JB. Nonmetric multidimensional scaling: A numerical method. Psychometrika. 1964;**29**:115-129

[7] Rao CR. The use and interpretation of principal component analysis in applied research. Sankhya. 1964;**A26**: 329-358

[8] Robert P, Escoufier Y. A unifying tool for linear multivariate statistical methods: The RV-coefficient. Applied Statistics. 1976;**A25**:257-265

[9] Mori Y, Tanaka T, Tarumi T. Principal component analysis based on a subset of qualitative variables. In: Hayashi C, editor. Data Science, Classification and Related Methods. Springer. 1997:547-554

Chapter 6

Space-Time-Parameter PCA for Data-Driven Modeling with Application to Bioengineering

Florian De Vuyst, Claire Dupont and Anne-Virginie Salsac

Abstract

Principal component analysis is a recognized powerful and practical method in statistics and data science. It can also be used in modeling as a dimensionality reduction tool to achieve low-order models of complex multiphysics or engineering systems. Model-order reduction (MOR) methodologies today are an important topic for engineering design and analysis. Design space exploration or accelerated numerical optimization for example are made easier by the use of reduced-order models. In this chapter, we will talk about the use of higher-order singular value decompositions (HOSVD) applied to spatiotemporal problems that are parameterized by a set of design variables or physical parameters. Here we consider a data-driven reduced order modeling based on a design of computer experiment: from high-dimensional computational results returned by high-fidelity solvers (e.g. finite element ones), the HOSVD allows us to determine spatial, time and parameters principal components. The dynamics of the system can then be retrieved by identifying the low-order discrete dynamical system. As application, we will consider the dynamics of deformable capsules flowing into microchannels. The study of such fluid-structure interaction problems is motivated by the use of microcapsules as innovative drug delivery carriers through blood vessels.

Keywords: HOSVD, spatio-temporal parametrized problem, approximation, reduced-order model, bioengineering, fluid-structure interaction, deformable capsules, dynamical system, machine learning, artificial intelligence

1. Introduction

Manufactured deformable microcapsules are intended to be used as drug carriers within the human vascular network to deliver drugs at specific targets (tumors, etc.). In order to design reliable capsules, one can make help of numerical simulation and high performance computing. The transportation of such capsules into microchannels is a three-dimensional fluid-structure interaction (FSI) problem involving a fluid flow within a confined environment and the deformation of hyperelastic membranes [1, 2].

The behavior of the capsule depends on dimensionless parameters such as the capillary number denoted by Ca and the aspect ratio a/ℓ between the capsule radius a and the channel characteristic length ℓ. The parameter vector $\boldsymbol{\mu} = (Ca, a/\ell)$, for which a capsule steady shape exists, lies in a bounded domain $D \subset \mathbb{R}^2$. We look for the time evolution of the capsule shape under a Lagrangian description. From an initial shape $\boldsymbol{X}$, we are interested in determining the capsule position $\boldsymbol{x}(\boldsymbol{X}, t, \boldsymbol{\mu})$ in the microchannel domain $\Omega \in \mathbb{R}^3$ at time t for a parameter vector $\boldsymbol{\mu} \in D$. By denoting $\boldsymbol{u}$ the displacement vector from the initial position, we have

$$\boldsymbol{x}(\boldsymbol{X}, \boldsymbol{\mu}, t) = \boldsymbol{X} + \boldsymbol{u}(\boldsymbol{X}, \boldsymbol{\mu}, t) \tag{1}$$

with $\boldsymbol{u}(\boldsymbol{X}, \boldsymbol{\mu}, 0) = \boldsymbol{0}$. The governing equations of the FSI problem include both kinematics and motion equations. At the membrane, we have equilibrium of the mechanical forces (mechanical equilibrium of the membrane and viscous stresses from the fluid). By denoting $\boldsymbol{v}$ the vector field of velocity at the membrane, the system of differential algebraic equations in abstract form reads

$$\dot{\boldsymbol{u}} = \boldsymbol{v}, \tag{2}$$

$$\boldsymbol{v} = \mathcal{F}_{\mu}(\boldsymbol{u}, \boldsymbol{X}). \tag{3}$$

Practically, there are different candidate computational approaches to discretize this system of equations. First, the initial capsule membrane has to be discretized by using a finite element triangular mesh made of nodes $\{\boldsymbol{X}_i\}_{i=1}^{I}$. Regarding time discretization, in [2], an explicit time scheme is used and the velocity field is determined by the use of a boundary integral method (BIM) coupled with a finite element method (FEM). A numerical stability condition imposes the use of small time steps. For a given parameter $\boldsymbol{\mu}$, the time evolution of capsule dynamics on the time intervals of interest generally requires hours of CPU time. To better understand the membrane behavior with respect to $\boldsymbol{\mu}$, a design of computer experiment (DoCE) is done: from a set of J parameter samples of $\boldsymbol{\mu}_j \in D, j = 1, \dots, J$, a spatio-temporal solution is computed for each $\boldsymbol{\mu}_j$, leading to a database of shape solutions under the form of a third-order tensor

$$\mathcal{T}_x = \left(\boldsymbol{x}\left(\boldsymbol{X}_i, \boldsymbol{\mu}_j, t^n\right)\right)_{i=1,\dots,I, j=1,\dots,J, n=0,\dots,N,} \in \mathbb{R}^{3I \times J \times (N+1)} \tag{4}$$

using a triangular finite element discretization of the membrane, a time discretization $t^n = n\Delta t$ (assuming that the time step is constant) and the parameter samples $\boldsymbol{\mu}_j$. Typically, for practical computations, $I = O(1000)$, $N = O(10000)$ and $J = O(100)$, so that the tensor database becomes rather huge (about $O(10)$ gigabytes). Of course, one can only store the solutions at coarser times steps and reduce N to $O(100)$ but the database remains rather big even in this case.

From this data tensor, one can imagine different use cases leading to different tools:

1. Data exploration and knowledge extraction;

2. Real-time rendering of capsule dynamics for better understanding;

3. Data-driven modeling of capsule dynamics in the whole parameter domain.

First and second items can be achieved by means of data dimensionality reduction. This leads to a lower storage of data in memory as well as a lower numerical complexity of processing and manipulation. In this chapter, we will consider a Higher Order Singular Value Decomposition (HOSVD), which is a generalization of Principal Analysis Component (PCA) to tensors. The third item involves a model-order reduction (MOR) methodology. From computed data, we would like to derive a lightweight dynamical system that reproduces the data and, even more, that is able to accurately estimate solutions for different parameter values $\boldsymbol{\mu}$. Data-driven model-order reduction first makes use of data-dimensionality reduction by a low-order tensor decomposition of the solutions according to some suitable spatial, temporal and parameter reduced bases (see [3]). In our application, this will give the truncated decomposition (where the solutions are here seen as functions):

$$\tilde{\boldsymbol{x}}(\boldsymbol{X},\boldsymbol{\mu},t) = \boldsymbol{X} + \sum_{k=1}^{K}\sum_{\ell=1}^{L}\sum_{m=1}^{M} a_{k\ell m}\,\boldsymbol{\varphi}_k(\boldsymbol{X})\,\psi_\ell(\boldsymbol{\mu})\,\omega_m(t) \tag{5}$$

for some expansion coefficients $a_{k\ell m}$ and some spatial functions $\boldsymbol{\varphi}_k$, parameter functions ψ_ℓ and temporal functions ω_m with $\omega_m(0) = 0$ ensuring $\boldsymbol{x}(\boldsymbol{X},\boldsymbol{\mu},0) = \boldsymbol{X}$. The truncation ranks K, L and M are expected to be small enough ($K \ll I$, $L \ll J$ and $M \ll N$). Discretized shape solutions returned by the Full-Order computational Model (FOM) are stored into a third-order tensor of data $\mathcal{T}_x$. Let $\tilde{\mathcal{T}}_x$ denote the truncated tensor expansion related to (5). It reads:

$$\tilde{\mathcal{T}}_x = A^0 \otimes \boldsymbol{e}^N + \sum_{k=1}^{K}\sum_{\ell=1}^{L}\sum_{m=1}^{M} a_{k\ell m}\,\boldsymbol{\Phi}_k \otimes \boldsymbol{\Psi}_\ell \otimes \mathbf{w}_m \approx \mathcal{T}_x \tag{6}$$

where $A^0 \in \mathbb{R}^{I\times J}$, $\left(A^0\right)_{ij} = \boldsymbol{X}_i$, $\boldsymbol{e}^N = (1, \dots, 1)^T \in \mathbb{R}^N$, $\boldsymbol{\Phi}_k \in \mathbb{R}^I$, $\boldsymbol{\Psi}_\ell \in \mathbb{R}^J$ and $\boldsymbol{w}_m \in \mathbb{R}^N$. In this chapter, we will seamlessly use the functional representation or the tensor one.

From this reduced form, we will then apply a kernel-based Dynamic Mode Decomposition (k-DMD, see [4]) to derive a dynamical system able to predict the capsule shape evolution over time for any value of the parameter $\boldsymbol{\mu}$. This will be developed in the next sections.

2. Higher-order singular value decomposition and truncation

2.1 Compact HOSVD

The higher-order singular value decomposition (HOSVD) of a tensor is a specific orthogonal Tucker decomposition. The classical computation of a HOSVD was introduced by L. R. Tucker [5] and further developed by L. De Lathauwer et al. [6]. Robust computations or improvements have been since proposed [6–8]. For a tensor $\mathcal{T}$ of order d, the idea is to compute the singular value decomposition of each factor-k flattening $T_{(k)}$ of a tensor $\mathcal{T}$, i.e. a "matricisation" of the tensor where the rows of the matrix are related to the k-th dimension.

In our case, we consider a third-order tensor and successive spatial, parameter and temporal flattening to find the singular vectors. The spatial tensor flattening $T_{(X)} \in \mathbb{R}^{I\times(J.N)}$ of the tensor $\left(\mathcal{T}_x - A^0 \otimes \boldsymbol{e}^N\right)$ is

$$\left(T_{(X)}\right)_{ip} = \boldsymbol{x}\left(\boldsymbol{X}_i, \boldsymbol{\mu}_j, t^n\right) - \boldsymbol{X}_i, \qquad p = j + (n-1)J \tag{7}$$

(meaning that the $\boldsymbol{\mu}$ and t dimensions are stacked in columns in the matrix). The SVD of $T_{(X)}$ provides r_x nonzero singular values with r_x corresponding singular orthonormal vectors $\boldsymbol{\Phi}_k$, $k = 1, \dots, r_x$. Similarly the parameter flattening leads to a rank r_μ with r_μ modes $\boldsymbol{\Psi}_\ell$, $\ell = 1, \dots, r_\mu$ and the time flattening a rank r_t with r_t modes $\boldsymbol{w}_m$, $m = 1, \dots, r_t$. The tuple $\left(r_X, r_\mu, r_t\right)$ is the multilinear rank of $\mathcal{T}_x$. Then tensor $\mathcal{T}_x$ can be written as

$$\mathcal{T}_x = A^0 \otimes \boldsymbol{e}^N + \sum_{k=1}^{r_X} \sum_{\ell=1}^{r_\mu} \sum_{m=1}^{r_t} a_{k\ell m} \boldsymbol{\Phi}_k \otimes \boldsymbol{\Psi}_\ell \otimes \boldsymbol{w}_m. \tag{8}$$

2.2 Approximation

Among the applications, HOSVD can be used to define a low-order approximation of tensors. The so-called truncated HOSVD [9–11] consists in truncating the expansion (8) at a given multilinear rank (K, L, M), $K \le r_X$, $L \le r_\mu$, $M \le r_t$, leading to (6). Let mlrank$(\mathcal{T})$ denote the multilinear rank of the tensor $\mathcal{T}$. It has been shown that the approximation (6) returns a quasi-optimal solution of the nonlinear non-convex least-square problem

$$\min_{\tilde{\mathcal{T}}} \quad \frac{1}{2} \|\mathcal{T}_x - \tilde{\mathcal{T}}\|_F^2 \tag{9}$$

(here $\|.\|_F$ is the Frobenius norm for tensors) subject to mlrank$\left(\tilde{\mathcal{T}}\right) = (K, L, M)$. The truncation ranks can be determined *a priori* according to the classical relative information content (RIC) of the SVD theory.

Eq. (6) can already be used as a compressed representation of the data, allowing for a lower storage complexity and a simpler manipulation, with low information loss if the RIC is high.

3. Reduced-order modeling of capsule dynamics

Eq. (6) provides a summarization of the family of spatio-temporal capsule shape solutions in the time interval $\left[t^0, t^N\right]$. Unfortunately, this algebraic model has no predictability capability for time $t > t^N$. To derive a predictable time-dependent model from the data tensor $\mathcal{T}_x$, one has to derive a differential system that approximates the FSI system of Eqs. (2) and (3). The HOSVD reduction can thus be valued in the context of model-order reduction.

Consider of parameter vector of interest $\boldsymbol{\mu} \in D$. The capsule position approximate solution (5) can be rewritten as

$$\tilde{\boldsymbol{x}}(\boldsymbol{X}, \boldsymbol{\mu}, t) = \boldsymbol{X} + \sum_{k=1}^{K} a_{\mu,k}(t) \varphi_k(\boldsymbol{X}) \tag{10}$$

with

$$a_{\mu,k}(t) = \sum_{\ell=1}^{L}\sum_{m=1}^{M} a_{k\ell m}\psi_\ell(\boldsymbol{\mu})\omega_m(t). \tag{11}$$

Let $\boldsymbol{a}_\mu(t)$ be the vector-valued function

$$\boldsymbol{a}_\mu(t) = \left(a_{\mu,1}(t), \dots, a_{\mu,K}(t)\right)^T \in \mathbb{R}^K. \tag{12}$$

We would like to derive a differential system of unknowns $\boldsymbol{a}_\mu(t)$. Since $\boldsymbol{x}(\boldsymbol{X},\boldsymbol{\mu},0) = \boldsymbol{X}$, we have the natural initial condition $\boldsymbol{a}_\mu(0) = \boldsymbol{0}$. Practically, from expansion (8), we can compute coefficients $a_{\mu,k}(t)$ at discrete times t^n, and have thus access to the list of coefficient vectors

$$\boldsymbol{a}_\mu^n = \boldsymbol{a}_\mu(t^n), \quad n = 0, \dots, N. \tag{13}$$

A dynamical reduced-order model consists in determining (or approximating) a Lipschitz continuous mapping $\boldsymbol{F}_\mu : \mathbb{R}^K \to \mathbb{R}^K$ such that

$$\boldsymbol{a}_\mu^0 = \boldsymbol{0}, \quad \boldsymbol{a}_\mu^{n+1} \approx \mathbf{F}_\mu\left(\boldsymbol{a}_\mu^n\right) \quad \forall n \in 0, \dots, N-1 \tag{14}$$

from the data $\left\{\boldsymbol{a}_\mu^n\right\}_{n=0}^{N}$. We get a low-order discrete dynamical system. Note that, here, we do not search for parameters of a model, but for the equations of the model themselves. Since the problem of finding such a mapping is infinite dimensional, one has to restrict the search to a mapping in a (suitable) finite dimensional functional space.

4. Koopman theory and dynamic mode decomposition

4.1 Koopman operator for discrete dynamical systems

Koopman theory is a powerful mathematical framework that re-expresses a general nonlinear discrete dynamical system as the knowledge of a linear (infinite dimensional) operator, the so-called Koopman operator or compositional operator. Today it is commonly used in machine learning and data-driven model-order reduction methodologies [4, 12]. Let us assume a discrete dynamical system in the form

$$\boldsymbol{a}^{n+1} = \boldsymbol{F}(\boldsymbol{a}^n), \quad n \in \mathbb{N} \tag{15}$$

for a Lipschitz continuous mapping $\boldsymbol{F}$ from $\mathbb{R}^d$ to $\mathbb{R}^d$. Let g be a function of a Banach space X, $g : \mathbb{R}^d \to \mathbb{R}$. So we have $g(\boldsymbol{a}^{n+1}) = (g \circ \boldsymbol{F})(\boldsymbol{a}^n)$. The Koopman operator related to $\boldsymbol{F}$ is defined as

$$\mathcal{K}g = g \circ \boldsymbol{F} \quad \forall g \in X. \tag{16}$$

Then we have

$$g\left(\boldsymbol{a}^{n+1}\right) = (\mathcal{K}g)(\boldsymbol{a}^n). \tag{17}$$

The knowledge of $\mathcal{K}$ includes the knowledge of $\boldsymbol{F}$. Indeed, by taking the particular observables $g_i(\boldsymbol{a}) = \boldsymbol{a} \cdot \boldsymbol{e}_i$, $i = 1, \dots, d$ where $\boldsymbol{e}_i$ is the i-th vector of the canonical basis of $\mathbb{R}^d$, we retrieve $(\boldsymbol{a}^{n+1})_i = F_i(\boldsymbol{a}^n)$, i.e. the i-th equation of (15). Of course, the linear Koopman operator acts on an infinite-dimensional functional space, so it is impossible to determine it exactly. However, one can search for an approximate Koopman operator $\tilde{\mathcal{K}}$ that acts on an approximate finite-dimensional space $\tilde{X} \subset X$.

The concept of (nonlinear) observables is to have a sufficiently large set of independent nonlinear functions of the state vector and measurements of them in order to identify the mapping $\boldsymbol{F}$. A natural question of interest is what are the best observables to choose. There is no absolute answer to this question, and the choice may depend on the underlying Physics. Without any a priori knowledge on the system of equations, one can use basis functions of a universal approximators like polynomials, Fourier or kernel-based functions for example.

4.2 Dynamic mode decomposition

The simplest choice of observables is the linear functions $g_i(\boldsymbol{a}) = \boldsymbol{a} \cdot \boldsymbol{e}_i$, $i = 1, \dots, d$. It leads to the search of a finite-dimensional approximation A of $\mathcal{T}$ from the full state vector data. The matrix A can be searched as the solution of the least square minimization problem

$$\min_{A \in \mathcal{M}_d(\mathbb{R})} \quad \frac{1}{2} \|Y - AX\|_F^2 \tag{18}$$

where $X = [\boldsymbol{a}^0, \boldsymbol{a}^1, \dots, \boldsymbol{a}^{N-1}]$ and $Y = [\boldsymbol{a}^1, \boldsymbol{a}^2, \dots, \boldsymbol{a}^N]$. Assuming $N \geq d$, the solution is given by $A = YX^T(XX^T)^{-1}$. The least square problem (18) can be eventually regularized for better conditioning by a Tykhonov regularization term [13, 14]. This practical approach of Koopman operator approximation is referred to as the dynamic mode decomposition (DMD) [4, 12]. This provides a linear dynamical model

$$\boldsymbol{a}^{n+1} = A\boldsymbol{a}^n \tag{19}$$

starting from a given initial condition $\boldsymbol{a}^0$. The solution of (19) which is $\boldsymbol{a}^n = A^n \boldsymbol{a}^0$ is bounded for any initial condition $\boldsymbol{a}^0$ as soon as $\rho(A) \leq 1$.

5. Kernel-based identification of dynamical systems

In the case of a strongly linear dynamical system, the linear model (19) can be not accurate enough. We have to include suitable nonlinear observables in the data and the model. In this section, observables are selected from kernel-based approximations [15]. Then we use the variant kernel-DMD (k-DMD, [4]) approach to identify $\boldsymbol{F}$.

A real-valued function k on $\mathbb{R}^d \times \mathbb{R}^d$ is called a positive definite kernel function if it is symmetric and if the following property holds:

$$\forall m \in \mathbb{N}^\star, \ \ \forall \{\boldsymbol{z}_i\}_{i=1}^m \in (\mathbb{R}^d)^m, \forall \{\alpha_i\}_{i=1}^m \in \mathbb{R}^m, \quad \sum_{i=1}^m \sum_{j=1}^m \alpha_i \alpha_j k(\boldsymbol{z}_i, \boldsymbol{z}_j) \geq 0. \tag{20}$$

In other words, the square matrix $\boldsymbol{K} = \left(k\left(\boldsymbol{z}_i, \boldsymbol{z}_j\right)\right)_{i=1,\dots,m, j=1,\dots,m}$ is positive semi-definite. A standard kernel function is for example the Gaussian one

$$k(\boldsymbol{z}, \boldsymbol{z}') = \exp\left(-\frac{1}{2}\frac{\|\boldsymbol{z}-\boldsymbol{z}'\|^2}{\sigma^2}\right) \tag{21}$$

for a given parameter $\sigma > 0$.

5.1 Kernel-based interpolation

Kernel functions can be used for interpolation in spaces of arbitrary dimension. Let $g : \mathbb{R}^d \to \mathbb{R}$ be a continuous function and assume that we know the values of $g(\boldsymbol{z}_i)$ at particular points $\boldsymbol{z}_i$, $i = 1, \dots, m$. Then one can define an interpolator $\mathcal{I}g$ of g defined as

$$\mathcal{I}g(\boldsymbol{z}) = \sum_{j=1}^{m} \alpha_j k\left(\boldsymbol{z}, \boldsymbol{z}_j\right) \tag{22}$$

where the coefficient vector $\boldsymbol{\alpha} = (\alpha_1, \dots, \alpha_m)^T$ is determined such that the interpolation property

$$\mathcal{I}g(\boldsymbol{z}_i) = g(\boldsymbol{z}_i) \quad \forall i \in \{1, \dots, m\}. \tag{23}$$

holds. The interpolation conditions clearly lead to the solution of the symmetric linear square system of size m

$$\boldsymbol{K\alpha} = \boldsymbol{b} \tag{24}$$

where $\boldsymbol{b} = (g(\boldsymbol{z}_1), \dots, g(\boldsymbol{z}_m))^T)$. Assuming that $\boldsymbol{K}$ is positive definite, then Eq. (24) has a unique solution. Let

$$k_i(\boldsymbol{z}) = k(\boldsymbol{z}, \boldsymbol{z}_i), \quad i = 1, \dots, m \tag{25}$$

and $V = \operatorname{span}(k_1, \dots, k_m)$. Considering any function $\tilde{g} \in V$, it is easy to check that $\tilde{g} = \mathcal{I}\tilde{g}$. One can derive the interpolation error

$$\begin{aligned} \|g - \mathcal{I}g\|_{L^\infty} &= \|g - \tilde{g} + \tilde{g} - \mathcal{I}g\|_{L^\infty} \\ &= \|g - \tilde{g} + \mathcal{I}\tilde{g} - \mathcal{I}g\|_{L^\infty} \\ &\leq \|g - \tilde{g}\|_{L^\infty} + |||\mathcal{I}||| \ \|g - \tilde{g}\|_{L^\infty} \end{aligned} \tag{26}$$

so that

$$\|g - \mathcal{I}g\|_{L^\infty} \leq (1 + |||\mathcal{I}|||) \inf_{\tilde{g} \in V} \|g - \tilde{g}\|_{L^\infty}. \tag{27}$$

The interpolation error is controlled by the best approximation error multiplied by a stability constant depending on the norm of the interpolation operator.

5.2 Use of kernel features and *k*-DMD

Let us go back to the parameterized dynamical system of interest (14) and consider a point cloud $\left\{\boldsymbol{a}_{(j)}\right\}_{j=1}^{m}$ of sample states in the admissible reduced state space $\mathcal{X} \subset \mathbb{R}^K$. The functions

$$k_j(\boldsymbol{a}) = k\left(\boldsymbol{a}, \boldsymbol{a}_{(j)}\right) \tag{28}$$

can be seen as features and thus be used as suitable nonlinear observables to approximate the Koopman operator. From any known full state vector $\boldsymbol{a}_\mu^n$ at time t^n, we build the vector of observables

$$\boldsymbol{\kappa}\left(\boldsymbol{a}_\mu^n\right) = \left(k_1\left(\boldsymbol{a}_\mu^n\right), k_2\left(\boldsymbol{a}_\mu^n\right), \dots, k_m\left(\boldsymbol{a}_\mu^n\right)\right)^T. \tag{29}$$

By definition of the Koopman operator, we have $k_i\left(\boldsymbol{a}_\mu^{n+1}\right) = (k_i \circ \boldsymbol{F})\left(\boldsymbol{a}_\mu^n\right) = (\mathcal{K}k_i)\left(\boldsymbol{a}_\mu^n\right)$. Then we look for a finite-dimensional approximation A_μ of the Koopman operator $\mathcal{K}$ in the sense

$$\boldsymbol{\kappa}\left(\boldsymbol{a}_\mu^{n+1}\right) \approx A_\mu \boldsymbol{\kappa}\left(\boldsymbol{a}_\mu^n\right) \quad \forall n \in \{0, \dots, N-1\} \tag{30}$$

The matrix A_μ is searched as the minimum of the least square problem

$$\min_{A \in \mathcal{MM}_m(\mathbb{R})} \quad \frac{1}{2} \|Y_\mu - A X_\mu\|_F^2 \tag{31}$$

where the entry and output data matrices are now $X_\mu = \left[\boldsymbol{\kappa}\left(\boldsymbol{a}_\mu^0\right), \boldsymbol{\kappa}\left(\boldsymbol{a}_\mu^1\right), \dots, \boldsymbol{\kappa}\left(\boldsymbol{a}_\mu^{N-1}\right)\right]$ and $Y_\mu = \left[\boldsymbol{\kappa}\left(\boldsymbol{a}_\mu^1\right), \boldsymbol{\kappa}\left(\boldsymbol{a}_\mu^2\right), \dots, \boldsymbol{\kappa}\left(\boldsymbol{a}_\mu^N\right)\right]$. We get the dynamical system $\boldsymbol{\kappa}\left(\boldsymbol{a}_\mu^{n+1}\right) = A_\mu \boldsymbol{\kappa}\left(\boldsymbol{a}_\mu^n\right)$ with a specified initial condition $\boldsymbol{\kappa}\left(\boldsymbol{a}_\mu^0\right)$.

Let us emphasize that the computational variables are now the $\boldsymbol{\kappa}\left(\boldsymbol{a}_\mu^n\right)$. But we still need the full state variables $\boldsymbol{a}^n$ to determine the displacements or the capsule shapes. The full state can be retrieved for example by interpolation: taking $g_i(\boldsymbol{a}) = \boldsymbol{a} \cdot \boldsymbol{e}_i$, we get

$$(\boldsymbol{a}_\mu)_i^{n+1} \approx \mathcal{I}g_i\left(\boldsymbol{a}_\mu^{n+1}\right) = \sum_{j=1}^{m} (\boldsymbol{\alpha}_j)_i k_i\left(\boldsymbol{a}_\mu^{n+1}\right) \tag{32}$$

with interpolation coefficients vectors $\{\boldsymbol{\alpha}_j\}_{j=1}^m$ such that

$$\boldsymbol{a}_{(i)} = \sum_{j=1}^{m} \boldsymbol{\alpha}_j k_j\left(\boldsymbol{a}_{(i)}\right) \quad \forall i \in \{1, \dots, m\} \tag{33}$$

(linear system of dimension $m \times K$). The coefficient vectors $\boldsymbol{\alpha}_j$ can be computed once for all.

5.3 Including the full state vector and the constants into the features

The k-DMD approach can be improved by including the full-state vector $\boldsymbol{a}_\mu$ itself into the input feature vector. Moreover, if the kernel functions are not able to perfectly reproduce the constant ones, for accuracy reasons it is justified to add the constants into the features. Considering the augmented vector $\boldsymbol{\eta}(\boldsymbol{a}_\mu)$ of observables

$$\boldsymbol{\eta}(\boldsymbol{a}_\mu) = \begin{bmatrix} \boldsymbol{a}_\mu \\ \boldsymbol{\kappa}(\boldsymbol{a}_\mu) \\ 1 \end{bmatrix} \in \mathbb{R}^{K+m+1}, \tag{34}$$

we look for a dynamical system in the form

$$\boldsymbol{a}_\mu^{n+1} = A_\mu \boldsymbol{\eta}\left(\boldsymbol{a}_\mu^n\right) \approx \boldsymbol{F}_\mu\left(\boldsymbol{a}_\mu^n\right) \tag{35}$$

with a constant rectangular matrix A_μ of size $K \times (K+m+1)$ to identify. The first advantage is that the output vector is the full state itself. The second one is the number of elements of A_μ which is less than m^2 as soon as $K(K+m+1) \leq m^2$. In this case, there are less coefficients to identify. The input and output matrices now are

$$X_\mu = \left[\boldsymbol{\eta}\left(\boldsymbol{a}_\mu^0\right), \boldsymbol{\eta}\left(\boldsymbol{a}_\mu^1\right), \dots, \boldsymbol{\eta}\left(\boldsymbol{a}_\mu^{N-1}\right)\right], \quad Y_\mu = \left[\boldsymbol{a}_\mu^1, \boldsymbol{a}_\mu^2, \dots, \boldsymbol{a}_\mu^N\right]. \tag{36}$$

Assuming $N \geq (K+m)$, the rectangular matrix A_μ solution of the least square problem

$$\min_{A \in \mathcal{M}_{K,K+m+1}(\mathbb{R})} \quad \frac{1}{2}\|Y_\mu - A X_\mu\|_F^2 \tag{37}$$

is computed as $A_\mu = Y_\mu X_\mu^T \left(X_\mu X_\mu^T\right)^{-1}$.

5.4 Summary and whole algorithm

We give a summary of the model-order reduction algorithm:

1. Input data: third-order tensor $\mathcal{T}_x$ (4) made of capsule shape solutions of size $3I \times J \times (N+1)$.

2. HOSVD + truncated approximation: compute (8) and get the truncated approximation with the truncated multilinear ranks (K, L, M):

$$\tilde{\mathcal{T}}_x = A^0 \otimes \boldsymbol{e}^N + \sum_{k=1}^{K} \sum_{\ell=1}^{L} \sum_{m=1}^{M} a_{k\ell m} \, \boldsymbol{\Phi}_k \otimes \boldsymbol{\Psi}_\ell \otimes \boldsymbol{w}_m \tag{38}$$

3. Online stage: choose a parameter vector $\boldsymbol{\mu}$. From (13), compute the data coefficients $\boldsymbol{a}_\mu(t^n) = \boldsymbol{a}_\mu^n$ (see Eq. (13)) from (11). Choose a kernel function $k(.,.)$, choose m and the centers $\boldsymbol{a}_{(j)}, j = 1, \dots, m$. Assemble the observables $\boldsymbol{\eta}\left(\boldsymbol{a}_\mu^n\right)$ and assemble the matrices X_μ and Y_μ (Eq. (36)). Compute the k-DMD matrix $A_\mu = Y_\mu X_\mu^T \left(X_\mu X_\mu^T\right)^{-1}$. Get the reduced-order dynamical system (35) with $\boldsymbol{a}_\mu^0 = \mathbf{0}$ as initial value.

6. Numerical results for capsule dynamics

The algorithm is applied to a problem of a deformable capsule flowing into a square-based microchannel (typical for microfluidic channels created by soft lithography) with a mean inflow velocity V [1, 2]. Related works and variant ROM approaches on this topic can be found in [14, 16]. The capsule dynamics is simulated by the full-order fluid–structure interaction solver Cite [1]. The fluid solver is based on the solution of the Stokes equations and a nonlinear Neo-Hookean law is used for the membrane. The initial capsule is spherical, corresponding to the shape at rest. The sphere is discretized with a finite element mesh made of $I = 2562$ vertices.

6.1 Study for a particular parameter couple $(Ca, \boldsymbol{a}/\boldsymbol{\ell})$

The objective of this subsection being to illustrate the methods on an example and show how to apply them, we only consider the snapshot FOM solutions for $(Ca, a/\ell) = (0.1, 0.9)$ for the sake of simplicity and brevity. **Figure 1** shows both membrane shapes and positions in the channel at different instants.

At each time $t^n = n\Delta t$, where the time step is equal to $\Delta t = 0.04$, a snapshot is saved and stored in the database. Note that all time quantities are non-dimensionalized by the factor ℓ/V. The resulting generated data matrix is used as the entry matrix for the ROM learning process. Then a truncated SVD is applied to get the spatial POD modes. In **Figure 2**, the four first eigenmodes Φ^k are plotted (more precisely this is a superposition of each mode onto the original spherical shape for a better visualization and understanding of their influence). Based on, the graphics $k \mapsto 1 - RIC(k)$ plotted in log scale in **Figure 3**, we decide to use a truncation rank K equal to $K = 10$, returning a relative information content of about $1 - 3.5 \times 10^{-5}$.

Then a reduced-order dynamical system for the capsule time evolution is searched. In this example, we compare two models: the first one is the affine approximation (denoted by ROM-A)

$$\boldsymbol{a}_\mu^{n+1} = A_\mu \boldsymbol{a}_\mu^n + \boldsymbol{b}_\mu \tag{39}$$

with the matrix A and the vector $\boldsymbol{b}$ to identify. It is equivalent to consider the vector of features $\boldsymbol{\eta}(\boldsymbol{a}) = (\boldsymbol{a}, 1)^T$. The second nonlinear model is built from the observable vector $\boldsymbol{\eta}(\boldsymbol{a}) = (\boldsymbol{a}, \boldsymbol{\kappa}(\boldsymbol{a}), 1)^T$ with the recurrent time scheme

$$\boldsymbol{a}_\mu^{n+1} = A_\mu \boldsymbol{\eta}\left(\boldsymbol{a}_\mu^n\right) \tag{40}$$

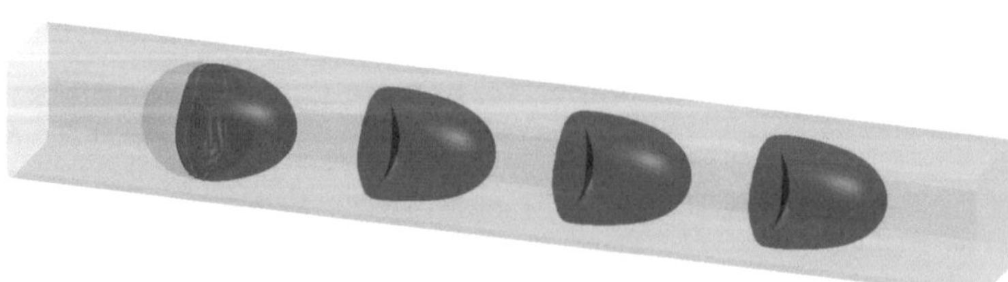

Figure 1.
Example of a microcapsule dynamics within a square-base channel for $(Ca, a/\ell) = (0.1, 0.90)$. Shapes and locations of the initially spherical capsule are shown at $t = 0$ (in transparency), 0.4, 2.8, 5.2, 7.6.

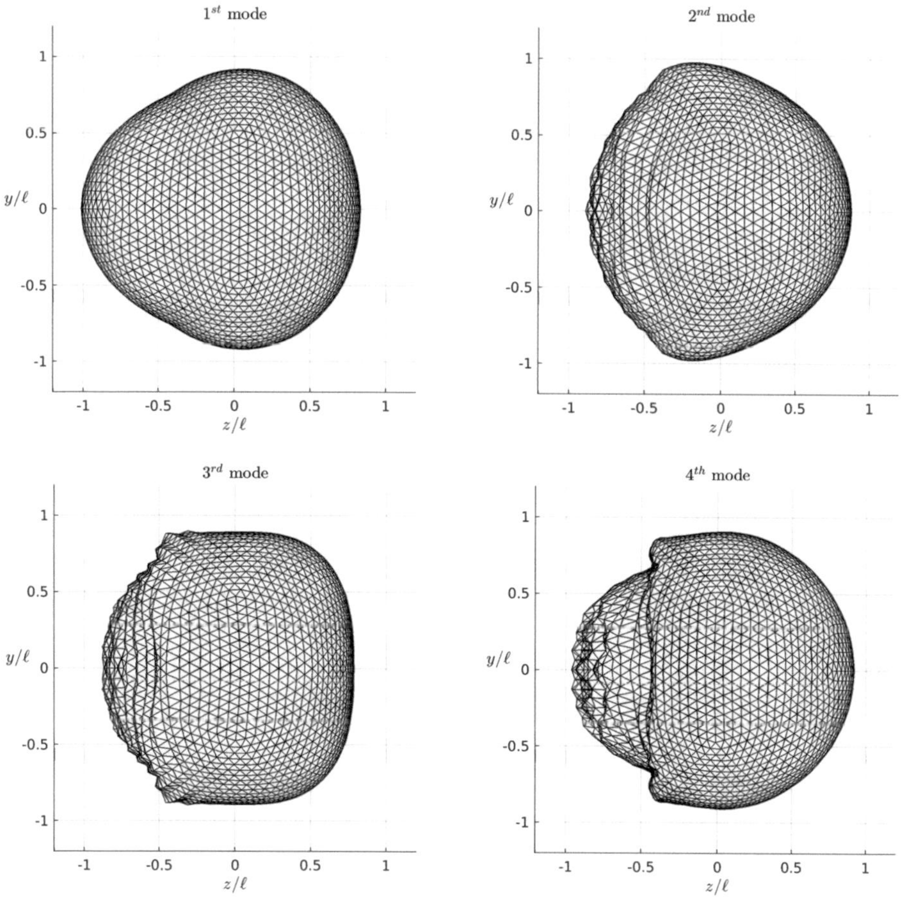

Figure 2.
SVD: Four first spatial principal components computed by the HOSVD. Each mode has been added on the initial spherical shape and amplified by a factor 2 for better visualization. Higher-order modes show oscillations at the rear of the capsule.

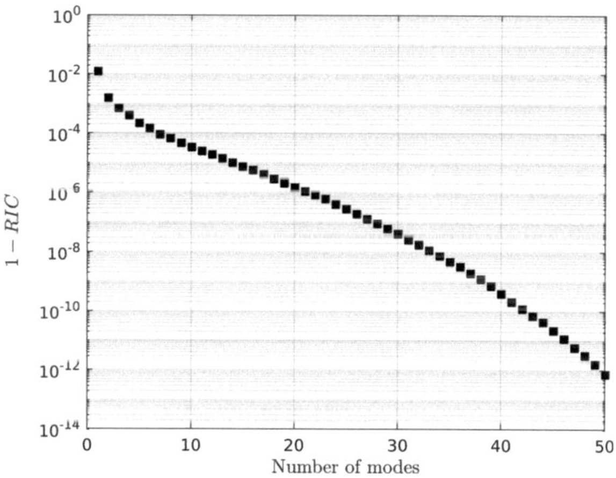

Figure 3.
SVD: Plot of $k \mapsto 1 - \mathrm{RIC}(k)$, *where* $\mathrm{RIC}(k)$ *represents the relative information content at truncation rank* k.

(ROM-B). For ROM-B, the Gaussian kernel function (21) is used. The standard deviation parameter σ in (21) is chosen as $\sigma = \max_n \|\boldsymbol{a}_\mu^n\| = \|\boldsymbol{a}_\mu^{N+1}\|_2$. In both cases ROM-A and ROM-B, the determination of the matrix A_μ by minimization of the least square problem (37) leads to a very small residual. In **Figure 4**, the logarithm of the relative time residual

$$n \mapsto \log_{10}\left(\|\boldsymbol{a}_\mu^{n+1} - A_\mu \boldsymbol{\eta}\left(\boldsymbol{a}_\mu^n\right)\|^2 / \|\boldsymbol{a}_\mu^{n+1}\|^2\right) \tag{41}$$

is plotted for each ROM model. One can observe values between 10^{-14} and 10^{-8}. The residual for ROM-B appears to be slightly smaller than that of ROM-A thanks to the added nonlinear terms. A surprising result is that the affine ROM-A model returns rather accurate results whereas the fluid–structure interaction problem is intrinsically nonlinear. In order to study the stability of the model ROM-A, in **Figure 5** we plot the complex eigenvalues of the square matrix A_μ. We observe that all the eigenvalues have a modulus less or equal to one. One of the eigenvalues is equal to 1 exactly up to Double Precision roundoff errors, meaning that there is a physical invariant in the system. It is known that the capsule volume is kept constant during time, because of the incompressibility of the fluid. For ROM-A, since $\boldsymbol{a}_\mu^{n+1} = A_\mu \boldsymbol{a}_\mu^n + \boldsymbol{b}$ and $\boldsymbol{a}_\mu^0 = \boldsymbol{0}$, we have

$$\boldsymbol{a}_\mu^n = A_\mu^n \boldsymbol{a}_\mu^0 + \sum_{k=0}^{n-1} A_\mu^k \boldsymbol{b} = \sum_{k=0}^{n-1} A_\mu^k \boldsymbol{b}. \tag{42}$$

The matrix A_μ is observed to be diagonalizable in $\mathbb{C}$. There is an invertible matrix P_μ such that $A_\mu = P_\mu \Lambda_\mu P_\mu^{-1}$ where Λ_μ is the diagonal matrix of the eigenvalues. Since it is observed that $\rho(\Lambda_\mu) = 1$, we have

$$\|\boldsymbol{a}_\mu^n\| \le \mathrm{Cond}(P_\mu)\, n \|\boldsymbol{b}\|, \quad \forall n \in \mathbb{N}, \tag{43}$$

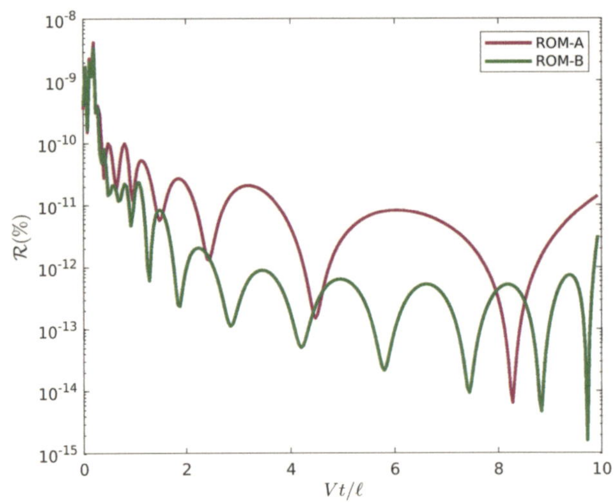

Figure 4.
Matrix identification. Log of the normalized residual $n \mapsto \log_{10}\left(\|\boldsymbol{a}_\mu^{n+1} - A_\mu \boldsymbol{\eta}\left(\boldsymbol{a}_\mu^n\right)\|^2 / \|\boldsymbol{a}_\mu^{n+1}\|^2\right)$ *for both ROM-A and ROM-B.*

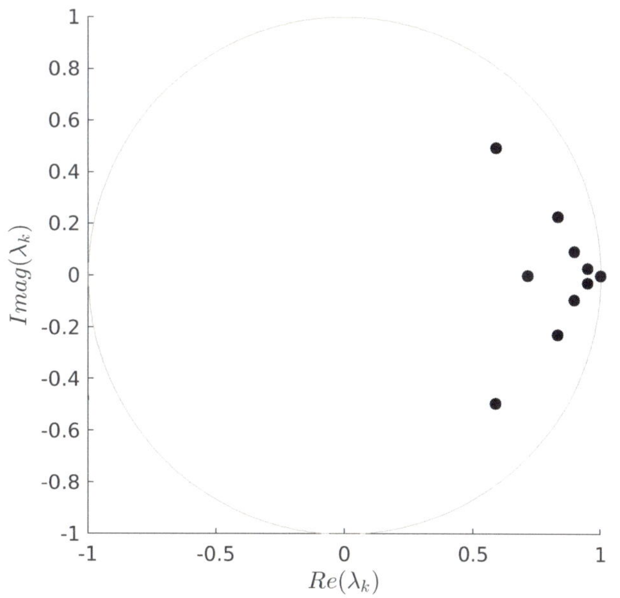

Figure 5.
Matrix identification. Eigenvalues of the computed matrix A_μ plotted in the complex plane for the ROM-A model. One of the eigenvalue is 1 up to round-off error.

showing that the coefficients in the PCA space grow at most linearly in time.

In **Figure 6**, we compare the computed capsule shapes and positions in the channel for the computed FOM capsules obtained at different times: $t = 0, 0.4, 2.8, 5.2$ and 7.6. What can be observed is that the ROM-B model returns very satisfactory results where the shape solutions fully overlap the FOM ones'at the eye norm'. Finely, we plot in **Figure 7** the time evolution of the errors in the capsule 3D shape of the ROM solutions as compared to the FOM solutions. The difference between the shapes are quantified by $\varepsilon_{\text{Shape}}(t)$, the ratio between the modified Hausdorff distance (MHD) computed between the FOM shape $\mathcal{S}_{\text{FOM}}(t)$ and the ROM shape $\mathcal{S}_{\text{ROM}}(t)$ and the channel characteristic length ℓ:

$$\varepsilon_{\text{Shape}}(t) = \frac{\text{MHD}(\mathcal{S}_{\text{FOM}}(t), \mathcal{S}_{\text{ROM}}(t))}{\ell}. \qquad (44)$$

The modified Hausdorff distance measures the maximum value of the mean distance between the two shapes to compare [17]. The ROM-A and ROM-B

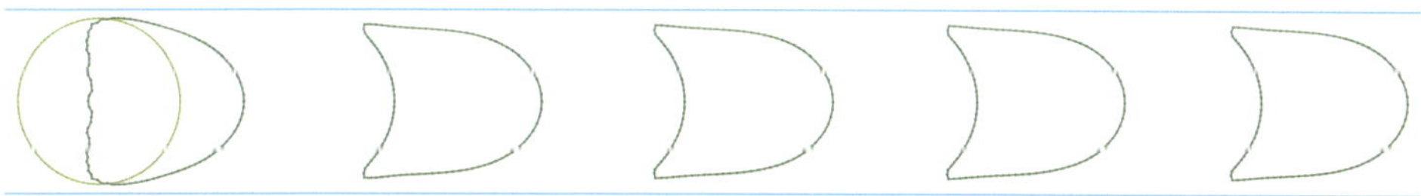

Figure 6.
Sequence of cross-section capsule shapes and positions in the microchannel from the initial spherical shape shown in light green at the beginning of the channel: Comparison of the FOM solutions (gray dots) and of the solutions computed from the dynamical k-DMD reduced-order model (dark green solid line) at the same instants as in ***Figure 1****:* $t = 0, 0.4, 2.8, 5.2, 7.6$.

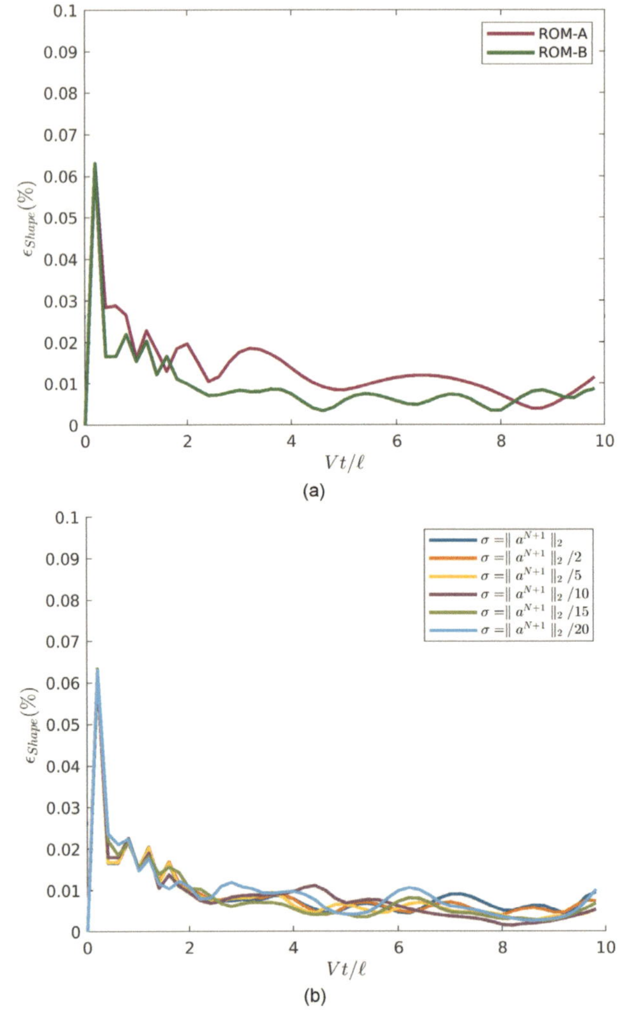

Figure 7.
(a) Comparison of the time evolution of the shape error between the affine DMD model (ROM-A with $K = 10$) and the kernel-based one (ROM-B with $K = 10$, $M = 5$). One can observe a maximum error less than 0.1% in both cases. (b) Sensitivity analysis of the parameter σ ($\|a^{N+1}\|_2 = 667.14$.)

models return very accurate solutions with maximum 0.1% error. It is also observed that the ROM-B models is slightly more accurate than the affine approximation.

6.2 HOSVD on the whole data tensor and error measurements on the whole database

Now the consider the whole database made of 55 samples in the parameter domain. In **Figure 8** we plot the location of the 55 chosen samples in the plane $(Ca, a/\ell)$. The design zone of interest corresponds to capsule shapes that can reach a steady state after a certain time.

A SVD is first performed on the μ-flattening of the data tensor $\mathcal{T}_x$. In **Figure 9**, we plot the four first parameter (normalized) eigenmodes in the parameter domain.

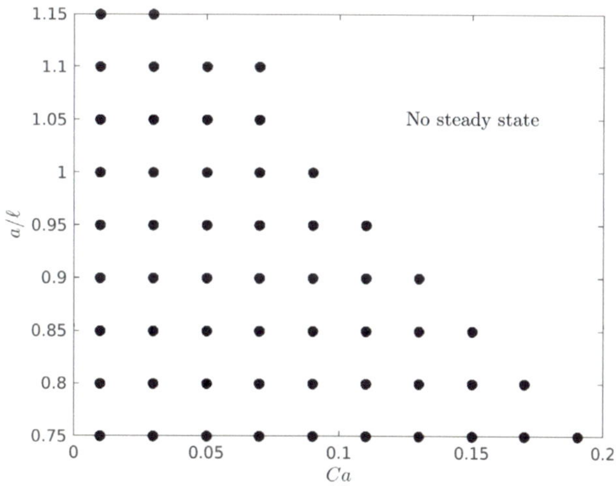

Figure 8.
Samples of the design of experiment in the parameter space $(Ca, a/\ell)$. The zone of interest corresponds to capsule shapes that reach a steady state after a certain time.

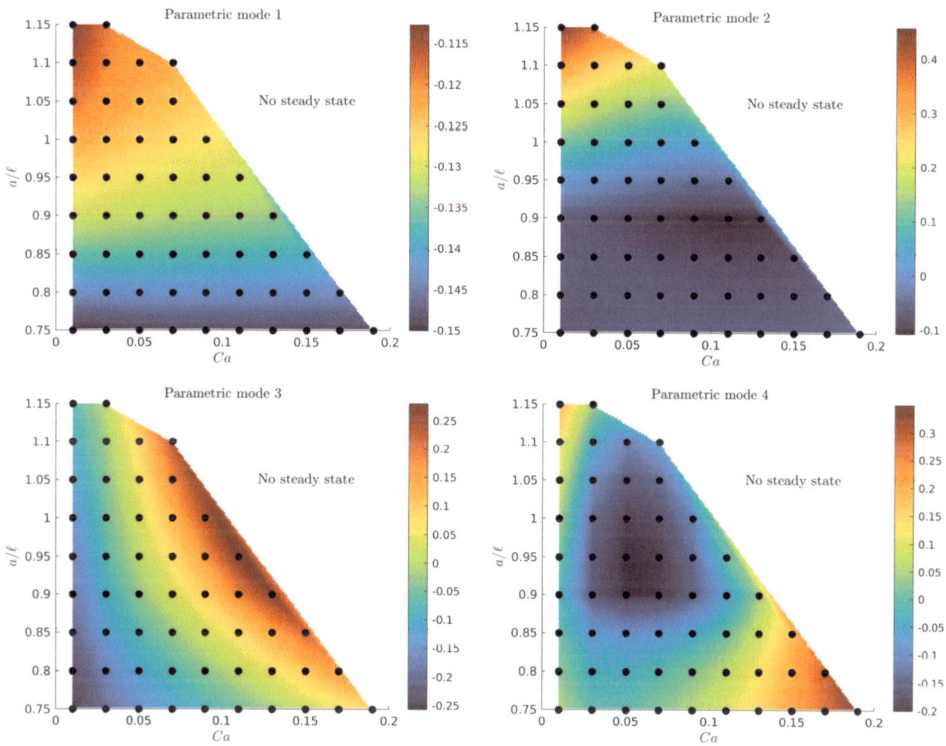

Figure 9.
HOSVD: The first four parameter eigenmodes in the parameter domain, computed from the μ-flattening of the data cube.

These modes give an idea on the dependency of the capsule shapes with respect to the parameters. To complete the analysis, we show in **Figure 10** the spectrum of the singular values for the μ-flattening matrix. One can observe a rather fast decay of the singular values especially for the ten first modes.

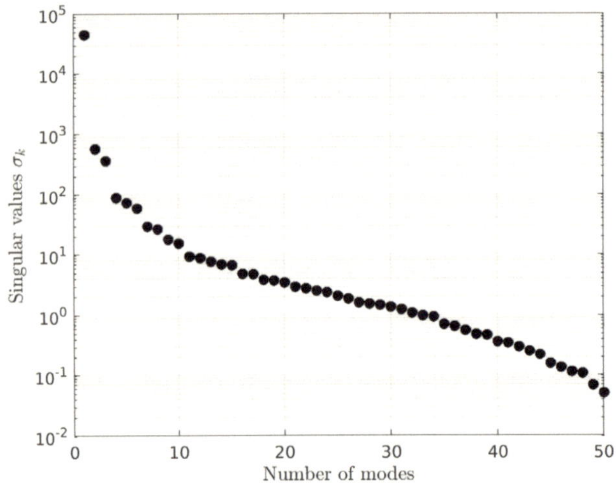

Figure 10.
Spectrum of the singular values of the μ-flattening matrix.

Next, we perform the SVD of the time t-flattening matrix of $\mathcal{T}_x$. The SVD provides us temporal eigenmodes. In **Figure 11**, the four first temporal eigenmodes ω_m, $m = 1, \dots, 4$ are plotted. The first one appears to be the linear function while the others add details especially in the transient zone of the capsule dynamics. The spectrum of the singular values again shows a fast decay especially for the six first modes.

To conclude this section, we have tested the accuracy of both ROM-A and ROM-B on the whole database. For each sample, we have derived a ROM model, i.e. a low-order dynamical system formulated in the PCA space. Then we have compared the ROM solution to the FOM solution by calculating $\varepsilon_{\text{Shape}}$ between the two capsule shapes. In **Figure 12**, the heat maps of $\varepsilon_{\text{Shape}}$ are plotted for ROM-A and ROM-B. One can observe a uniform accuracy over the whole parameter domain, with errors less than 0.1%, thus showing the efficiency of the approach. Reported computational speedups are between 5000 and 10,000 using ROM models. A computer with two Intel Xeon GOLD 6130 CPU (2.1 Ghz) has been used for the numerical tests.

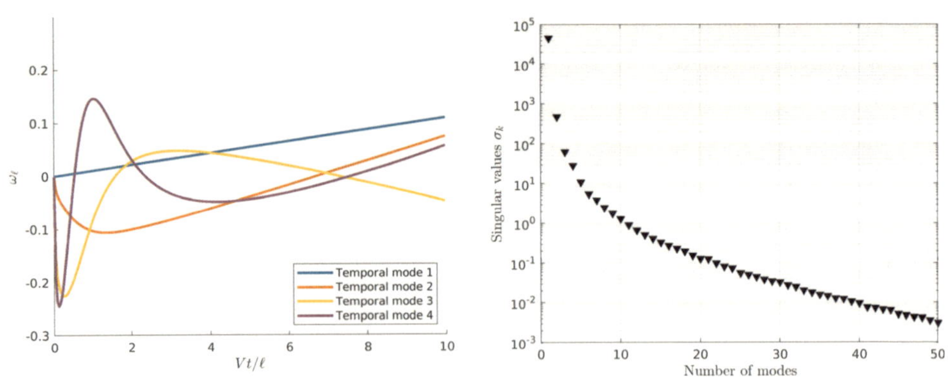

Figure 11.
HOSVD. Left: First four temporal eigenmodes computed from the SVD decomposition of the t-flattening data tensor. Right: Spectrum of the singular values.

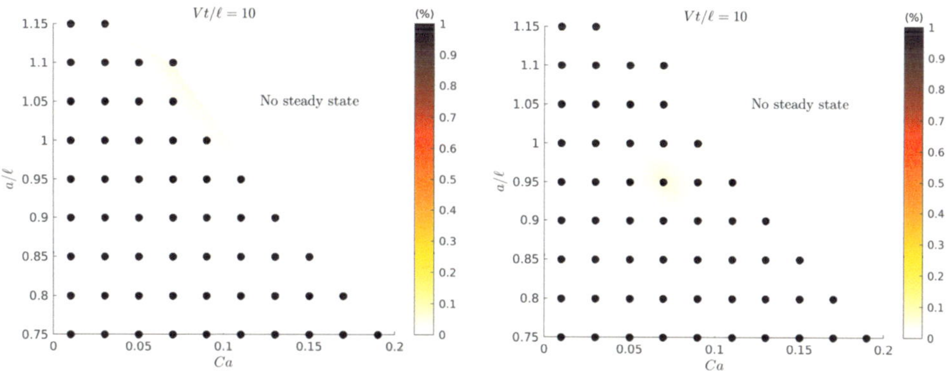

Figure 12.
Heat maps of the modified Hausdorff distance between the FOM solutions and the ROM ones at dimensionless time $Vt/\ell = 10$. Left: ROM-A, right: ROM-B. Errors are less than 0.1%.

7. Discussion

Having shown how to apply the affine DMD and k-DMD models and the very high precision in prediction that they offer, we now would like to provide some further comments and remarks on model order reduction and DMD-type approaches.

7.1 Kolmogorov n-width

The method is efficient if the spectrum of singular values decays rapidly, leading to a small truncation rank K. If the spectrum decays slowly, there are two possible reasons for that: either the entropy (variety) of information in the data is high, or the solutions do not live in a linear space but rather on a nonlinear manifold. To fix the problem, one can proceed by performing a preliminary clustering of the data, scanning the parameter dimension. One can either use standard clustering techniques such as K-means, or a multidimensional scaling (MDS) approach. Then for each cluster, one can consider again a HOSVD and reduced-order approach suitable for each element of the cluster.

7.2 Selection of the kernel functions and kernel interpolation points

As already mentioned, the choice of the kernel function depends on the applications, on the behavior of solutions and/or on the underlying Physics. Without any a priori information, one can use universal approximation kernels like the Gaussian one. The accuracy of the results will also strongly depends on the choice of the kernel interpolation points $\boldsymbol{a}_{(j)}$. The sampling $\{\boldsymbol{a}_{(j)}\}_{j=1}^{m}$ has to correctly fill in the admissible space, or at least the state-space trajectory of interest. There are different possible strategies. A first candidate is the use of a clustering approach applied to the state-space data. The points $\boldsymbol{a}_{(j)}$ are then the centroids of each cluster. But one can consider more sophisticated approaches like a greedy iterative approach that controls the interpolation error on the data. At each iteration an interpolation point $\boldsymbol{a}_{(j)}$ is added at the location of worst interpolation error, considering all the sample solutions.

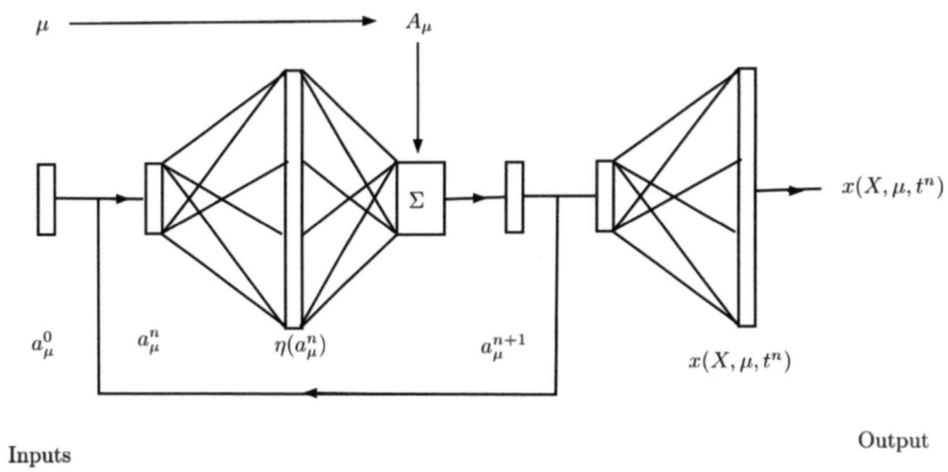

Figure 13.
Interpretation of the method as a recurrent neural network (RNN) in the PCA space.

7.3 Interpretation in terms of a recurrent neural network

Let us remark that the approach can be reinterpreted as a (supervized) two-layer recurrent artificial neural network (RNN) (**Figure 13**) [18]. The first layer consists in generating the features $k_i(\boldsymbol{a}_\mu)$. The second layer is a linear matrix–vector multiplication using the matrix A_μ.

8. Conclusions

In this chapter, the higher-order singular value decomposition has been proved to be a flexible and valuable tool in the data-driven reduced-order modeling of solutions of space–time-parameter problems, which are today at the heart of many industrial applications. The methodology has been tested on a problem of fluid–structure interaction of a deformable microcapsule flowing into a microchannel. Stokes equations have been used in the fluid whereas a nonlinear hypereleastic law has been used for the membrane. Different shape solutions computed by the full-order model have been stored into a third-order tensor. First, HOSVD allows us to compute spatial, temporal and parameter principal components and at the same time to compress the data. We get a low-order representation of the solutions with a shared spatial reduced basis. Spatial principal components are observed to provide suitable details in the shape solutions. The modes are arranged in decreasing order of importance according to the relative information content criterion. Next, additional ingredients such as kernel approximation and kernel-based dynamic mode decomposition are used to determine a reduced-order dynamical system for any parameter vector in the admissible parameter domain. The resulting low-dynamical system can be seen as an encoded recurrent neural network set into the PCA space. The approach allows us to explore the different shape solutions and visualize their evolution in the channel in real time.

Acknowledgements

The authors would like to thank Prof. Pierre Villon (UTC) for fruitful discussions on this topic. The project received funding from the European Research Council (ERC) under the European Union's Horizon 2020 Research and Innovation Programme (Grant agreement No. ERC-2017-COG - MultiphysMicroCaps).

Conflict of interest

The authors declare no conflict of interest.

Nomenclature

SVD	Singular value decomposition
HOSVD	Higher-order singular value decomposition
PCA	Principal component analysis
MOR	Model order reduction
FOM	Full order model
ROM	Reduced order model
FSI	Fluid–structure interaction
DoCE	Design of computer experiment
POD	Proper orthogonal decomposition
RIC	Relative information content
DMD	Dynamic mode decomposition
k-DMD	Kernel-based dynamic mode decomposition
MHD	Modified Hausdorff distance
RNN	Recurrent neural network

Author details

Florian De Vuyst[1*†], Claire Dupont[2†] and Anne-Virginie Salsac[2†]

1 Compiègne Applied Mathematics Laboratory, University of Technology of Compiègne – CNRS, Compiègne, France

2 Biomechanics and Bioengineering Laboratory (UMR CNRS 7338), University of Technology of Compiègne – CNRS, Compiègne, France

*Address all correspondence to: florian.de-vuyst@utc.fr

† These authors contributed equally.

References

[1] Hu X-Q, Salsac A-V, Barthès-Biesel D. Flow of a spherical capsule in a pore with circular or square cross-section. Journal of Fluid Mechanics. 2012;**705**: 176-194. DOI: 10.1017/jfm.2011.462

[2] Walter J, Salsac A-V, Barthès-Biesel D, Le Tallec P. Coupling of finite element and boundary integral methods for a capsule in a stokes flow: Numerical methods for a capsule in a stokes flow. International Journal for Numerical Methods in Engineering. 2010;**83**: 829-850. DOI: 10.1002/nme.2859

[3] Audouze C, De Vuyst F, Nair PB. Reduced-order modeling of parameterized PDEs using time–space-parameter principal component analysis. International Journal for Numerical Methods in Engineering. 2009;**80**: 1025-1057

[4] Kutz JN, Brunton SL, Brunton BW, Proctor JL. Dynamic Mode Decomposition: Data-driven Modeling of Complex Systems, Society for Industrial and Applied Mathematics. Philadelphia, PA: SIAM; 2016. p. 234. DOI: 10.1137/1.9781611974508

[5] Tucker LR. Some mathematical notes on three-mode factor analysis. Psychometrika. 1966;**31**(3):279-311. DOI: 10.1007/bf02289464

[6] De Lathauwer L, De Moor B, Vandewalle J. A multilinear singular value decomposition. SIAM Journal on Matrix Analysis and Applications. 2000; **21**(4):1253-1278. DOI: 10.1137/s0895479896305696

[7] Hackbusch W. Tensor spaces and numerical tensor calculus. In: Springer Series in Computational Mathematics. Vol. 42. Berlin, Heidelberg: Springer; 2012. p. 605. DOI: 10.1007/978-3-642-28027-6

[8] Grasedyck L. Hierarchical singular value decomposition of tensors. SIAM Journal on Matrix Analysis and Applications. 2010;**31**(4):2029-2054. DOI: 10.1137/090764189

[9] Vannieuwenhoven N, Vandebril R, Meerbergen K. A new truncation strategy for the higher-order singular value decomposition. SIAM Journal on Scientific Computing. 2012;**34**(2): A1027-A1052. DOI: 10.1137/110836067

[10] Goldfarb D, Zhiwei Q. Robust low-rank tensor recovery: Models and algorithms. SIAM Journal on Matrix Analysis and Applications. 2014;**35**(1): 225 253. DOI: 10.1137/130905010

[11] Eldén L, Savas B. A Newton–Grassmann method for computing the best multilinear rank - (r_1, r_2, r_3) approximation of a tensor. SIAM Journal on Matrix Analysis and Applications. 2009; **31**(2):248-271. DOI: 10.1137/070688316

[12] Brunton SL, Budišić M, Kaiser E, Kutz JN. Modern Koopman theory for dynamical systems. ArXiv. 2021

[13] Benner P, Himpe C, Mitchell T. On reduced input-output dynamic mode decomposition. Advances in Computational Mathematics. 2018;**44**: 1751-1768

[14] Dupont C, De Vuyst F, Salsac AV. Data-driven kinematics-consistent model order reduction of fluid-structure interaction problems: Application to deformable microcapsules in a stokes flow. Journal of Fluid Mechanics. Under revision. ArXiv Preprint. DOI: 10.48550/arXiv.2203.13725

[15] Fasshauer G, McCourt M. Kernel-based approximation methods using

MATLAB. Interdisciplinary Mathematical Sciences. 2015;**19**:536. DOI: 10.1142/9335

[16] Boubehziz T, Quesada-Granja C, Dupont C, Villon P, De Vuyst F, Salsac A-V. A data-driven space-time-parameter reduced-order model with manifold learning for coupled problems: Application to deformable capsules flowing in microchannels. Entropy. 2021;**23**:1193. DOI: 10.3390/e23091193

[17] Dubuisson M, Jain AK. A modified Hausdorff distance for object matching. Proceedings of 12th International Conference on Pattern Recognition. 1994;**1**:566-568

[18] Tsoi AC, Back A. Discrete time recurrent neural network architectures: A unifying review. Neurocomputing. 1997;**15**(3–4):183-223. DOI: 10.1016/S0925-2312(97)00161-6

Chapter 7

Principal Component Analysis in Financial Data Science

Stefana Janićijević, Vule Mizdraković and Maja Kljajić

Abstract

Numerous methods exist aimed at examining patterns in structured and unstructured financial data. Applications of these methods include fraud detection, risk management, credit allocation, assessment of the risk of default, customer analytics, trading prediction, and many others, creating a broad field of research named Financial data science. A problem within the field that remains significantly under-researched, yet very important, is that of differentiating between the three major types of business activities—merchandising, manufacturing, and service based on the structured data available in financial reports. It can be argued that, due to the inherent idiosyncrasies of the three types of business activities, methods for assessment of the risk of default, methods for credit allocation, and methods for fraud detection would all see an improved performance if reliable information on the percentage of entities' business activities allocated to the three major activities would be available. To this end, in this paper, we propose a clustering procedure that relies on Principal Component Analysis (PCA) for dimensionality reduction and feature selection. The procedure is presented using a large empirical data set comprising complete financial reports for various business entities operating in the Republic in Serbia, that pertain to the reporting period 2019.

Keywords: data science, principal component analysis, random forest algorithm, financial data, financial reporting

1. Introduction

The established financial reporting system within an entity is the basic source of information on its financial position and results. The economic and financial globalization of the world market has emphasized the importance of high quality financial reporting. For the business decision making process, financial and audit reports are the main source of information, as they contain information on financial position, business results, changes in equity, cash-flows and other reliable information [1]. Development of the capital market and the increase in the number of interested parties (investors) created even higher demand of reliable, on time and fair financial statements as the main results of financial reporting. The regulation of the relationship between the state and society, owners of capital and management, various stakeholders and society, and others; has been further improved by a quality financial reporting and audit process. However, in order to fulfill their main purpose for all interested parties, financial statements must provide information that is true, objective, comprehensible, comparable and uniform [2]. In the first place, financial statements have to be publicly available, which is usually regulated by law. For example, Law on Accounting of the Republic of Serbia prescribes that all business entities have to submit their financial reports to the competent institution which later publishes them on the official internet site [3]. Information contained in financial statements can be used for numerous purposes. For example, other business entities can use them in the process of making business, financial, investment and other decisions. Likewise, banks and financial institutions can use them in order to approve loans or assess investment risks related to the certain business entity. However, financial information contained in financial statements are not processed and represent a raw data that should be analyzed in order to assess the performance of a certain business entity. Aside Notes to financial statements, as one of the qualitative statements that business entities prepare and report, all other statements are quantitative in nature and offer hundreds of pieces of data. Therefore, it is of great importance to perform certain type of analysis on the collected data in order to gain a solid basis for business decision making process. Analysis of financial statements is one of the most common methods of assessing business performance. The main goal of conducting the analysis of financial statements is to obtain information on the performance of the observed company, i.e. liquidity, profitability and solvency. Measuring financial performance using compiled and disclosed financial statements is a quantitative analysis of the position of the observed company, including the way in which the company uses the capital invested in business. High quality analysis of the performance of the observed entity provides a comprehensive image of the business, including meeting the information needs of stakeholders. The authors [4] point out in their paper that the analysis of financial performance is crucial in determining the efficiency in terms of the use of available resources. Likewise, an entity owners will be able to assess management skills and decisions that have been made in previous, as well as in current reporting period, so that they could analyze entities strengths, weaknesses and therefore improve their overall performance [5–7].

Some pieces of data disclosed in financial statements have informational power to be used on their own, such as Total assets, Sales revenue, or Net result. However, informational power of data increases when they are put into relation with other pieces of data. Therefore, financial statements analysis using ratios has been one of the most commonly used methods of assessing business performance. Financial ratio is a relative magnitude of two (or more) selected numerical values taken from financial statements. For example, relation between Net result and Equity will provide information on how much dollars of profit an entity earns for each dollar invested in equity. Results of financial statements analysis can be used to compare performance of a certain entity over a period of time, or for comparison with other entities within the industry. However, since financial statements analysis takes time and there are numerous financial ratios that analysts could use (and the fact that most of these ratios are correlated), the number of ratios that are being calculated and assessed should be reduced so that an analyst could focus on several of them without losing data that could be relevant for the analysis [8]. One of the methods that can be used is Principal Component Analysis (PCA), which reduces number of observed variables for any further, regression, or any other type of analysis [9]. PCA analysis has found its numerous purposes in different industries, for example, in image compressing [9–11], as well as in biometrics or "bioimaging" where physical characteristics of a person are used for its identification with application on communication devices and security systems.

The significance of PCA results is reflected in the fact that they can be used for more effective and efficient analysis of performance of certain entity, or for all business entities within a certain industry, or if analyzed financial data is related to whole economy, than results could be used for the analysis of all entities within it. The main advantages of PCA are precision of results; reduction of time needed for the analysis and evaluation of results; as well as reduction of related costs and efforts of the analyst.

With the development of technology, we have gained the ability to generate massive amounts of data. The use of correct methodologies for data analysis has become essential when dealing with complex financial challenges. In this paper, we discuss the theory underlying PCA. This type of analysis is one of the most used statistical tools in the field of financial data analysis. To ensure that the proper method is used for the analysis, theoretical knowledge and an comprehension of statistical methods are essential.

1.1 General postulates of PCA

PCA is primarily designed as a statistical technique that selectively reduces the dimensionality of data in complex data sets while preserving maximum variance. Since research in the financial sector involves both a large amount of data and a large number of variables simultaneously, it is difficult for us to perform analysis for this type of data.

Visualization techniques are only useful in two or three dimensional spaces, and single-variable analysis does not provide precise results due to overlapping variance. To achieve dimensionality reduction, it is necessary to generate principal components, i.e., a new set of variables containing a linear combination of the original variables. PCA can be used for a variety of tasks. A very small number of components are sufficient to cope with the variability of a data set. Since the number of components is reduced by using principal components, the complexity of the analysis itself is also reduced by avoiding analyzing a large number of output variables.

The standard PCA procedure takes as its starting point a data set in which m numerical variables are observed for each n individuals. These data are defined by the vectors $x_1, \dots, x_m$ or $n \times m$ of the data matrix X. The jth column is the vector x_j resulting from the jth variable. Linear combinations of columns for an X matrix with maximum variance are calculated as $\sum_{j=1}^{m} c_j x_j = Xc$. Here c stands for the vector of constants $c_1, c_2, \dots c_m$. The variants of such a linear combination are obtained as $var(Xc) = c'Mc$. Here M stands for an exemplary covariance matrix. Finding a linear combination with maximum variance is the same as finding a m dimensional vector c that maximizes the quadratic form $c'Mc$. For this reason, it is necessary to enter another constraint, which is usually unit norm vectors. Such vectors require $c'c = 1$. This problem is the same as maximizing $c'Mc - \lambda(c'c - 1)$, where λ represents the Lagrange multiplier. Equating it to the zero vector gives the following equation:

$$Mc - \lambda c = 0 \Leftrightarrow Mc = \lambda c \qquad (1)$$

This equation is valid even when the eigenvectors are multiplied by -1. Here, c is the eigenvector and λ is the corresponding eigenvalue for the covariance matrix M. We need the largest λ_1, the largest eigenvalue, and the corresponding eigenvector c_1. Eigenvalues are defined by the corresponding eigenvector $c : var(Xc) = c'Ma = \lambda c'c = \lambda$. The covariance matrix M is a symmetric $m \times m$ matrix and has exactly m real eigenvalues. $\lambda_k(k = 1, \dots, m)$ can be defined together with the corresponding eigenvectors to form a set of vectors that are orthonormal. An example of this is $c'_m c_m = 1$ if $m = m'$. The eigenvectors of M are used to obtain up to m linear combinations of $Xc_k = \sum_{j=1}^{m} c_{jk} x_j$ that maximize the variances. The fact that the covariance between the two linear combinations of Xc_k and $Xc_{k'}$ is obtained from $c'_k M c_k = \lambda_k c'_k c_k = 0$ if $k' \neq k$, leads to results of uncorrelatedness [12]. Linear combinations of Xc_k represent the principal component of a data set. There are several PCA terms used for specific values. Elements of linear combinations Xc_k are called principal component scores (PCA scores) and eigenvectors c_k are also called principal component loads (PCA loads). These contain a generic element $x^*_{ij} = x_{ij} - \overline{x}_j$, where x^*_j represents the observed value for variable j.

The $n \times m$ matrix labeled X^* contains columns with centered variables x^*_j, resulting in the following equation:

$$(n-1)M = X^{*'}X^* \tag{2}$$

1.2 Premises of PCA

For the final outcome of the PCA assessment to be successful and significant, numerous conditions must be met. Initially, it is crucial that the data entered are uninterrupted and that variables should be measured on an interval or ratio scale. This condition must be met because PCA tests important correlation patterns for these variables.

Another crucial requirement is that the relationships between the individual pairs of variables are linear. If there are nonlinear relationships between the individual pairs of variables, appropriate data transformation techniques, such as logarithmic transformations, should be considered. Presumptions for PCA are filling missing values with not null values, outliers handling, and normalization scaling. All outliers should be filtered out prior to analysis, as they can bias the results by affecting the magnitude of the correlation.

To obtain more accurate estimates for the correlation population parameters, a large sample size is required. The data sets must be linear in order to be formed. The basic principle of PCA is that high variance must be taken into account, while variables with lower variance can be considered noise and are not taken into account. All variables must be processed at the same level of measurement.

1.3 Features extraction in PCA

Eq. (2) associates the eigenvalue decomposition of the covariance matrix M and the singular value decomposition of the matrix X^* with the centered column data. For dimension $n \times m$ and rank r, where it must be $r \leq min\,\{n, m\}$, the matrix Y can be calculated as follows:

$$Y = ULA' \tag{3}$$

Where U and A represent the matrices $n \times r$ and $m \times r$ containing orthonormal columns $U'U = I_r = A'A$, where I_r represents the identity matrix $r \times r$. L is the $r \times r$ diagonal matrix. The columns A are also called right singular vectors and represent eigenvectors for the $m \times m$ matrix $Y'Y$ associated with its non-zero eigenvalues. Columns U are also called left singular vectors and represent eigenvectors for the $n \times n$ matrix YY' associated with its non-zero eigenvalues. Singular values of Y represent diagonal elements of the matrix, denoted by L. These elements are non-negative square roots for the non-zero eigenvalues of the two matrices $Y'Y$ and YY'. We consider that the diagonal elements are sorted from the largest to the smallest element, which determines the order of the columns U and A, except for singular values that are equal [12]. This is true in all cases except when the singular values are equal. If we assume that $Y = X^*$, then the right singular vectors for the matrix X^* are vectors c_k of principal component loads. Because of the orthogonality of columns A, columns $X^*A = ULA'A = UL$ are the principal components for X^*. The types of these principal components are obtained by squaring the singular values of X^* and dividing by $n - 1$. This results in the following equation:

$$(n-1)M = X^{*'}X^* = (ULA')'(ULA') = ALU'ULA\prime' = AL^2A' \tag{4}$$

Here L^2 stands for a diagonal matrix with one square of the singular values. With this equation we get the eigenvalue decomposition for the matrix $(n-1)M$. The singular value decomposition for the X^* matrix with the data centered in the column is equivalent to PCA. Taking the rank r in the matrix Y, which has the magnitude $n \times m$, the matrix Y_q, which has the same magnitude but the second rank $q < R$ and whose elements reduce the sum of squared differences with the corresponding elements of Y, is obtained as:

$$Y_q = U_q L_q A'_q \tag{5}$$

Here L_q stands for the diagonal matrix of dimensions $q \times q$, which contains the first largest diagonal element q of L and U_q. A_q stands for the matrices $n \times q$ and $m \times q$ obtained by keeping the q columns in U and A. The number of rows n from the rank r of the matrix X^* defines the scatter plot from the number n of points in the r dimensional subspace $\mathbb{R}^m$, where the beginning of the gravity center for the scatter plot is located. It follows that the best approximation of the n points in this scatterplot in the q dimensional subspace, obtained by using X_q^* rows, is given by this equation. That means that the sum of the squared distances between the given points in each scatterplot is minimal, as in Pearson's original approach [13]. The q axis system defines the main subspace. It can be concluded that PCA is a dimensionality reduction method where a set of m original variables can be replaced by a given set of q variables. In the case of $q = 2$ or $Q = 3$, it is possible to make a graphical approximation for n points in the scatter plot, and it is very often used to visualize the whole data set. A very important aspect is that the results are incremental in their dimensions.

The variability associated with the set of retained principal components can be used to ensure the quality of any q dimensional approximation. The trace, i.e. the sum of the diagonal elements, of the covariance matrix M is equal to the sum of the variances of the m variables. It is possible to achieve this with the help of matrix theory results. It is easy to prove that this number is also the sum of the variances of all m principal components. Consequently, the proportion of the overall variation accounted for by a given principal component is a standard measurement of its quality and it's equal to:

$$\pi_j = \frac{\lambda_j}{\sum_{j=1}^{m} \lambda_j} = \frac{\lambda_j}{tr(M)'} \tag{6}$$

The trace of M is labeled $tr(M)$. Due to the incremental behavior of principal components, we can speak of a proportion of the total variance explained by a set M of principal components, which is usually expressed as a percentage of the total variance and is accounted for:

$$\sum_{j \in M} \pi_j \times 100\% \tag{7}$$

It is a common approach to use a pre-specified percentage of the total variance to determine how many principal components to keep, but graphical constraints often lead to keeping only the first two or three principal components. The percentage of total variance is a basic tool for measuring the quality of these low-dimensional graphical representations of the data set.

The biggest problem is the number of components needed to obtain a sufficient number of variances while achieving a reduction in dimensionality. There are several ways to determine the components, and one of them is to set a threshold.

The next very popular approach is the "Scree Plot" [14], where the components are arranged on the X-axis from largest to smallest with respect to their eigenvalues. In this way, we can see a very large difference between important and less important components. The only drawback to this approach is that it is subjective in determining the correct number of components.

The most popular method is parallel analysis [15], where PCA is performed with as many variables as the original data set includes. The average eigenvalues between the original data set and the simulated data set are measured. Any values from the original data that are lower than the data in the simulated set are discarded.

1.4 Sparse PCA

PCA has many advantages. In terms of maximizing variance in Q dimensions, PCA provides the best possible representation of a m dimensional data set in q dimensions $q < m$. However, the new variables it defines are often linear functions of all the m original variables, which is a downside. Multiple variables with not so simple coefficients are common for larger m, making the components difficult to read. A number of PCA adjustments have been proposed to facilitate interpretation of the q dimensions while limiting the loss of variance that results from not using the principal components themselves. There is a compromise between interpretability and variance. Two types of adjustments are briefly outlined below.

Factor analysis is a method that is often combined with PCA and it inspires the concept of rotating principal components [16]. Assume that A_q is the $m \times q$ matrix whose columns are the loadings of the first q of the principal components. Then XA_q is the $n \times qnq$ matrix whose columns are the scores of the first q of the principal components for the n observations. Let us assume that T is an orthogonal $q \times q$ matrix. Multiplying A_q by T causes orthogonal rotation of the axes within the space spanned by the first q of principal components, resulting in $B_q = A_q T$, a $m \times q$ matrix whose columns are the charges of the q rotated principal components. XB_q is an $n \times q$ matrix containing the associated values of the rotated principal components. Any orthogonal matrix T can be used to rotate the components, but it is preferable to make the rotated components easy to understand. For this reason, T is chosen to maximize simplicity. A variety of such criteria have been proposed, some of which involve non-orthogonal rotation. The criterion where an orthogonal matrix T is chosen for maximizing $Q = \sum_{k=1}^{q} \left[\sum_{j=1}^{m} b_{jk}^4 - \left(\frac{1}{m}\right) \left(\sum_{j=1}^{m} b_{jk}^2 \right)^2 \right]$, where b_{jk} is the $(j, k)^{th}$ member of B_q, is probably the most commonly used. No variance is lost when considering the rotated q dimensional space, since the sum of the variances of the q rotated components is the same as the sum of the variances of the unrotated components. Successive maximization of the non-rotated principal components is lost, which means that the sum of the variances of the q rotated components is the same as the sum of the variances of the non-rotated components. A disadvantage of rotation is the necessary choice between different rotation criteria, although this choice often makes less difference than the choice of the number of components to rotate. If q is increased by 1, the rotated components may look substantially different. That is because this does not happen in principal components with defined non-rotated nature.

Another method of simplifying the principal components is to limit the charges of the new variables. This is called adding a constraint. There are several variants of this strategy, one of which uses LASSO linear regression [17], that represents least absolute shrinkage and selection operator. In this approach, SCoTLASS components are discovered, solving the same optimization problem as PCA, but with the additional constraint $\sum_{j=1}^{m} |c_{jk}| \leq \tau$, where tuning parameter is τ. The constraint has no effect for $\tau > \sqrt{m}$, and principal components are generated; however, more charges are pushed to zero at a lower value, which simplifies the interpretation. These simplified components must have less variation than the corresponding number of principal components, and multiple values of τ are often examined to find a reasonable compromise between added simplicity and loss of variance. One distinction between rotation and constraint techniques is that the second has the advantage that some loadings in linear functions are set exactly to zero for interpretation, whereas this is usually not the case with rotation. Sparse variants of PCA are type of adjustments in which many coefficients are zero, and numerous studies of such principal components have been conducted in recent years. Hastie et al. [18] provides a good overview of this work.

1.5 Robust PCA

PCA is inherently sensitive to the occurrence of outliers and thus to large errors in data sets [19]. As a result, efforts have been made to define robust variants of PCA, and the terminology RPCA has been used to refer to several approaches to this problem. Huber's early work focused on robust alternatives to covariance or correlation matrices and how they could be used to generate robust principal components [20]. The demand for methods to process very large data sets sparked renewed interest in robust PCA variants. This led to PCA research lines, especially in areas such as machine learning, image processing, web data analysis, and many others.

Wright et al. [21] defined RPCA as the sum of two $n \times m$ components, a low-rank component L and a sparse component S in an $n \times m$ data matrix X. Identifying the matrix components of $X = L + S$ that minimize a linear combination of two separate component norms was defined as a convex optimization task and calculated as:

$$\min_{L,S} \|L\|_* + \lambda\|S\|_1 \tag{8}$$

where $\|L\|_* = \sum_r \sigma_r(L)$ is the nuclear norm of L, and $\lambda\|S\|_1 = \sum_i \sum_j |s_{ij}|$ is the l_1 norm of matrix S.

2. Related work

PCA was first introduced into mechanics by [22], as an analogue of the axis theorem. It was later named "PCA" by [23]. The range of applications in finance and economics is extensive. Take as an example [24], who used PCA to document three factor structures. Stock and Watson [25] used PCA to monitor economic development and activity, as well as the inflation index. Egloff et al. [26] used PCA as a way to analyze the dimensions of inconsistent dynamics. Volatility is a statistical measure that can be used to determine these inconsistencies using a two-factor volatility model. This includes long-term and short-term fluctuations in the volatility structure. Baker and Wurgler [27] used PCA to measure investors sentiment, i.e., their positive or negative view. This was done according to the principle of the number of sentiment proxies before Baker, [28] created the policy uncertainty index. This index represents potential risks in the near future.

The most important item in the construction of PCA is the estimation of the eigenvalues of the covariance matrix sample. Anderson and Weeks [29] and Anderson [30] showed that sample eigenvalues were consistent when dealing with asymptomatic sentiment proxy results. Waternaux [31] proved that similar results are obtained with simple eigenvalues as long as there is a fourth moment in the data. In addition to the discussions in the [32] book, [33] was able to establish the asymptotic distribution of eigenvectors using generalized assumptions.

However, this PCA approach to eigenvalues has some downsides. The first problem is certainly dimensionality, which can be noticed when the cross sectional dimension grows simultaneously with the sample in the same period. Then inconsistencies occur. Another problem arises from linear data types that do not include nonlinear patterns. A third problem [34] arises from the dependence of the asymptotic theory on fixed assumptions for the analysis. For these reasons, we have a problem when we use PCA for reimbursement data. Most of the time, we need years of data to make an assumption, which in turn leads to other problems, such as permanence and consistency of non-fixed parameters. This type of data has backlogs and volatility times often vary.

These problems stimulate the improvement in this field and motivate the development of tools for PCA methods. The approach to the problem, where the number of occurances grows in fixed time periods, touches all the listed downsides. Theoretically, it is known that as the frequency of the sample increases, the estimated variance and covariance increase. This is true until the microstructure of the market begins to take effect. Incidentally, this is not a serious problem if we choose a sampling frequency of minutes, which we use as opposed to the below one second time interval most often used for liquid stocks. A high frequency asymptotic analysis with the cross-sectional dimension is expected as the time interval increases sharply. This high frequency asymptotic framework allows us to perform non-parametric analysis as well as independent, non-static and analysis without underlying parameters as is the case with low frequency processes.

Asymptotic theory is very common in many contexts. Jacod et al. [13] and Jacod and Podolskij [35] also dealt with one problem that we deal with in this paper, where the cross sectional dimensions are invariant and the process is continuous. Mykland and Zhang [36] designed an alternative theory to the one put forward by [37], that discuss inference for volatility function dependence. It is based on the aggregation of local estimates and uses a finite number of blocks. Saha et al. [38] considered the expected values of the integrated covariance matrix under conditions where there is an error measure and the matrix is large containing high frequency data. Tao et al. [39] addressed work on the convergence rate. Jacod and Rosenbaum [40] analyzed estimators, composed of aggregating functions of estimates. They did so using integrated quarticity estimation. Heinrich and Podolskij [41] discussed empirical covariate matrices of Brownian integrals. Here is discussed the measurement of the leverage effect and its evaluation by the integrated correlation method [42].

PCA analysis can be used in analysis of financial data for different purposes. For example [43] used it to identify the type of impact on grouped impact factors, such as assessing the quality of accounting information and facilitating the process of financial analysis conducted by different users. On the other hand, [44] used PCA to assess the impact of the evolution of Finnish standards on IFRS (International Financial Reporting Standards). Finally [45] used PCA analysis to determine the macroeconomic impact on the profitability of Romanian listed companies, using data from 1997 to 2007, and identified following indicators: liquidity, solvency, and firm's dimension.

When it comes to the use of PCA analysis in financial statements analysis, four papers that focus on Romanian listed companies will be reviewed first. All papers emphasize the importance of using PCA analysis in the analysis of key financial ratios. In the first paper author [46] analyzed the data of 16 initial variables which he grouped into 3 new variables (general efficiency indicator, indicator in correlation with historical debts of companies and development indicator (given long-term debt and deferred income). Those three variables where able to explain 96.72% of initial variability. In the second paper, [47] analyzed data for 2010 including initially seven indicators of standard financial analysis and they reduced them to only two (which explain 94% of initial variability). In third paper, [48] used data from the stock exchange in the period 2006–2011 to identify the main components of financial statements which explain 79.08% of initial variability. The same group of indicators has been used by [43] on research sample that consisted of 111 companies from Madrid stock exchange and 32 companies from Eurostoxx50 for reporting periods 2005–2007. Research results showed that those six indicators explained 87% of total variance, with the first two indicators at app 44% of total variance.

3. Case study—PCA and cluster analysis in financial accounting data

3.1 Research methodology

In order to provide an answer on defined research question, 3.013 medium and large business entities were selected by random and used as a research sample. Financial statements for 2019 reporting period have been downloaded manually from the official website of the Business Registers Agency (BRA). BRA is a state administrative body that collects financial statements and corresponding audit reports of business entities that operate within the territory of the Republic of Serbia. Information published by BRA is used for financial analysis of business entities and as a basis of decision-making process. Afterwards, data from the pdf files containing financial statements have been copied and recorded in pre-set up tables in Excel files. Namely, medium and large business entities in the Republic of Serbia have an obligation to prepare and disclose full set of financial statements, consisting of balance sheet, income statement, cash-flow statement, statement of changes in equity and notes to financial statements. Since all previously mentioned statement, except notes to financial statements, are quantitative in nature, they were used for this research. Values originally disclosed in RSD, as the reporting currency, were converted into euros by using the average exchange rate of euros on the balance sheet date (31st December). Values of each financial statement line is presented in thousands, and therefore they are presented as such in this research [49].

Financial statement item lines in official financial statements are marked by corresponding automatic data processing number (in Serbian: Automatska obrada podataka—AOP), that belongs to the national nomenclature system. These markings are used in order to perform control of mathematical calculations before each financial statement is accepted for publishing by BRA. They also serve as an instrument of connecting data and information regarding the same financial statement item presented in financial statements. Balance sheet items cover automatic data processing numbers from 0001 to 0465; income statement from 1001 to 1071; statement of cash-flows from 3001 to 3047; and statement of changes in equity from 4001 to 4252. **Table 1** shows the formulas used for the calculation of the selected financial indicators that will be used in this research. Having in mind that these variables will be used in order to differentiate business entities to three major types of business activities, these variables have been selected by a common sense.

Variables	Derived from
Fixed assets in total assets	AOP2/AOP71
Percent sales of merchandise in total operating revenue	AOP1002/AOP1001
Percent sales of products and services in total operating revenue	AOP1009/AOP1018
Percent cost of merchandise sold in total operating expenses	AOP1019/AOP1018
Percent cost of material in total operating expenses	AOP1023/AOP1018
Percent fuel and energy cost in total operating expenses	AOP1024/AOP1018
Percent wage cost in total operating expenses	AOP1025/AOP1018
Percent productive service cost in total operating expenses	AOP1026/AOP1018
Percent depreciation cost in total operating expenses	AOP1027/AOP1018
Percent raw material in total assets	AOP45/AOP71
Percent WIP in total assets	AOP46/AOP71
Percent finished products in total assets	AOP47/AOP71
Percent WIP and finished products in total assets	(AOP46 + AOP47)/AOP71
Percent merchandise in total assets	AOP48/AOP71

Table 1.
Calculation of selected financial indicators.

3.2 Algorithm

Data preparation is a key process in data analysis. The basic preparation and cleaning procedures are:

- Preparing a copy of the table
- Adding new attributes
- Conversion of column types
- General data cleaning and adjustment

Specifically, the cleaning includes the following items:

- Editing date variables—the most common formatting problems
- Recoding of zeros/missing values
- Decoding categorical variables using labels and hot encoding
- Arranging outliers
- Application of normalization/standardization/ log transformation
- Calculating descriptive statistics—mean, median, mode, standard deviation, variance, rank, etc.
- Calculating inferential statistics - distributions, t-value, p-value, frequencies, cross-tabulations, correlation, covariance, etc.

More advanced techniques include:

- Coding:

Categorical variables are labeled as character variables and must be converted to a factor type for modeling purposes. Queues perform this task.

- Outliers:

For numeric variables, we can identify deviations numerically by the value of the bias.

- Normalization/logarithmic transformation:

One of the techniques to normalize the biased distribution is logarithmic transformation. First, a new variable is created, while later the value of the bias of this new variable is calculated and printed.

- Standardization:

One of the standardization techniques is that all characteristics are centered around zero and have approximately the variance of one unit. Scaling is used so that the variable is converted. The result is that these variables are standardized with a mean of zero.

As part of the preparation for PCA, firstly missing values from the dataset were filled with zeros. After that, the data was scaled by using a standard scaler, which standardizes features by removing the mean and scaling to unit variance. The preprocessed dataset, was then used for:

- PCA
- Sparse PCA
- Robust PCA

All three of the PCA methods were instanciated with the number of components set to 7. After PCA, the now transformed data went through several clustering methods for the purpose of comparing results. The clustering methods that were used for each PCA are:

- K-means clustering
- Agglomerative clustering
- BIRCH clustering
- Gaussian Mixture
- Spectral clustering

Furthermore, each of the clustering methods were executed with just the preprocessed data, without PCA, also for the purpose of comparing results.

Algorithm 1: Principal Component Analysis.

---

procedure:
    Data preparation: $X \leftarrow X^*$
    Compute dot product matrix: $X^{*'}X^* \leftarrow (n-1)M$
    Eigenanalysis: $AL^2A' \leftarrow X^{*'}X^*$
    Compute eigenvectors: $U \leftarrow X^*AL$
    Keep first 7 components: $U_7 \leftarrow [u_1 \cdots u_7]$
    Compute 7 features: $Y \leftarrow U_d{}'X$
end procedure.

---

4. Results and discussion

4.1 Comparative results—total variance explained

This chapter discusses the outcomes of PCA and cluster analysis. The initial variables that load on the principal components are studied. Correlations or covariances between the original variables and the principal components correlate with the loadings. The variable loadings are contained in a loading matrix, which is created by multiplying the eigenvector matrix by a diagonal matrix containing the square root of each eigenvalue. The entries are determined by the component extraction method used. Non-standardized loadings show the covariance between mean-centered variables and standardized component values, regardless of whether the extraction is based on the singular value decomposition of the matrix or the eigenvalue decomposition of the covariance matrix.

The eigenvalue decomposition of the correlation matrix results in the standardized charges. The correlations between the original variables and the component scores are represented by these loadings. Because they always vary between −1 and 1 and are independent of the scale used, standardized charges are easy to read. In most cases, a threshold is set and only variables with loadings above this threshold are examined.

The total variance presents sum of variances of principal components. The ratio between the variance of principal component and the total variance is the fraction of variance explained by a principal component.

Figure 1 shows total variance explained by using three methods of PCA. The steepest increase belongs to the PCA line, which cumulative explained variance is app. 87%. This line is almost parallel to the line from Sparse PCA which cumulative explained variance is 83%. However, when it comes to Robust PCA line it has been noticed that cumulative explained variance is only app. 26% and the increase of values is minimal.

PCA: The highest fraction of explained variance among these variables is 32%, and the lowest one is 5%. Cumulative explained variance is 86% (see **Table 2**).

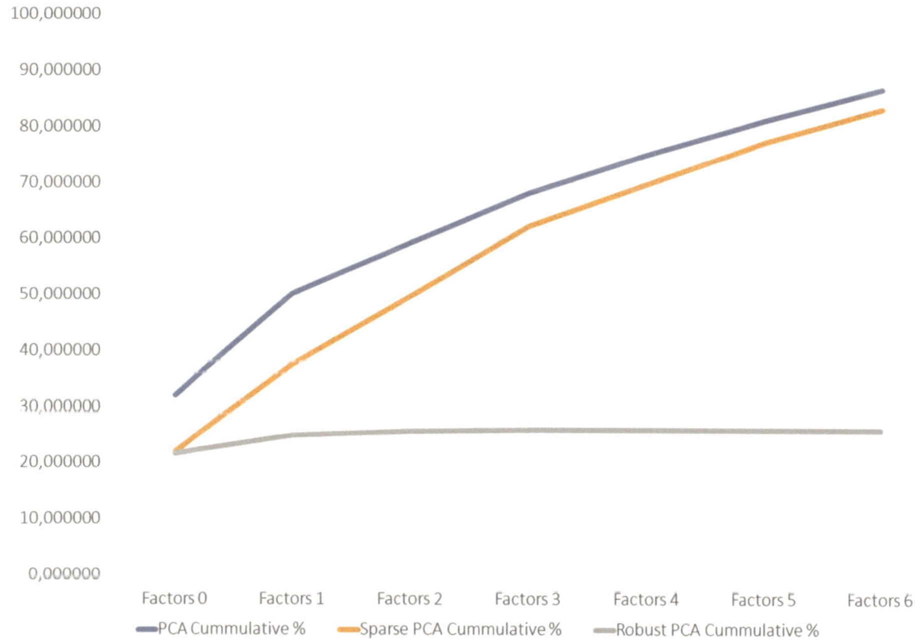

Figure 1.
Total variance explained.

Factors	Total	% of variance	Cumulative %
Factor 0	4.491515	32.082248	32.082248
Factor 1	2.540717	18.147978	50.230226
Factor 2	1.269778	9.069843	59.300069
Factor 3	1.243867	8.884762	68.184831
Factor 4	0.961330	6.866641	75.051473
Factor 5	0.867145	6.193891	81.245364
Factor 6	0.760536	5.432398	86.677761

Table 2.
PCA total variance explained.

Sparse PCA: The highest fraction of explained variance among these variables is 21%, and the lowest one is 5%. For instance, variables together explain 83% of the total variance (see **Table 3**).

Robust PCA: The highest fraction of explained variance among these variables is 21%, and the lowest one is 0%. For instance, variables together explain 25% of the total variance (see **Table 4**).

PCA is the best approach for this kind of data, regarding number of features.

Factors	Total	% of variance	Cumulative %
Factor 0	3.078591	21.989939	21.989939
Factor 1	2.186255	15.616108	37.606047
Factor 2	1.698036	12.128828	49.734874
Factor 3	1.757003	12.550022	62.284897
Factor 4	1.047037	7.478832	69.763729
Factor 5	1.062211	7.587224	77.350953
Factor 6	0.809469	5.781923	83.132875

Table 3.
Sparse PCA total variance explained.

Factors	Total	% of variance	Cumulative %
Factor 0	3.035926	21.685184	21.685184
Factor 1	0.454951	3.249650	24.934834
Factor 2	0.108168	0.772628	25.707462
Factor 3	0.020284	0.144884	25.852346
Factor 4	0.006630	0.047355	25.899701
Factor 5	0.000018	0.000128	25.899829
Factor 6	0.000000	0.000000	25.899829

Table 4.
Robust PCA total variance explained.

4.2 Communalities

The amount of variance in each variable considered is represented by the communalities. The variance in each variable explained by all components or factors is estimated using the initial communalities.

The percent fuel and energy cost in total operating expenses is given here with 88% variance. The percent productive service cost in total operating expenses is given here with 75% variance. The percent finished products in total assets here is 75% of the estimated variance (see **Table 5**).

The percent fuel and energy cost in total operating expenses here is 91% variance. The percent finished products in total assets here is 80% of the estimated variance. The percent productive service cost in total operating expenses here is 74% variance (see **Table 6**).

Columns	Communality
Percent merchandise in total assets	0.159427
Percent sales of merchandise in total operating revenue	0.222216
Percent cost of merchandise sold in total operating expenses	0.224299
Percent sales of products and services in total operating revenue	0.236318
Fixed assets in total assets	0.347415
Percent cost of material in total operating expenses	0.411423
Percent raw material in total assets	0.426201
Percent WIP and finished products in total assets	0.449704
Percent depreciation cost in total operating expenses	0.683213
Percent wage cost in total operating expenses	0.729997
Percent WIP in total assets	0.731771
Percent finished products in total assets	0.745349
Percent productive service cost in total operating expenses	0.752027
Percent fuel and energy cost in total operating expenses	0.880639

Table 5.
PCA communalities.

Columns	Communality
Percent merchandise in total assets	0.191833
Percent sales of products and services in total operating revenue	0.227810
Percent sales of merchandise in total operating revenue	0.260545
Percent cost of merchandise sold in total operating expenses	0.263888
Fixed assets in total assets	0.354743
Percent cost of material in total operating expenses	0.407825
Percent raw material in total assets	0.417451
Percent WIP and finished products in total assets	0.451553
Percent depreciation cost in total operating expenses	0.555661

Columns	Communality
Percent wage cost in total operating expenses	0.695148
Percent WIP in total assets	0.719447
Percent productive service cost in total operating expenses	0.742714
Percent finished products in total assets	0.800108
Percent fuel and energy cost in total operating expenses	0.911274

Table 6.
Sparse PCA communalities.

Columns	Communality
Percent WIP in total assets	0.200472
Percent merchandise in total assets	0.317793
Percent finished products in total assets	0.333984
Percent depreciation cost in total operating expenses	0.345393
Percent fuel and energy cost in total operating expenses	0.349862
Percent sales of products and services in total operating revenue	0.365996
Percent raw material in total assets	0.433737
Percent WIP and finished products in total assets	0.444081
Percent cost of material in total operating expenses	0.519423
Fixed assets in total assets	0.651365
Percent productive service cost in total operating expenses	0.680299
Percent cost of merchandise sold in total operating expenses	0.745842
Percent sales of merchandise in total operating revenue	0.789024
Percent wage cost in total operating expenses	0.822730

Table 7.
Robust PCA communalities.

The percent wage cost in total operating expenses here is 82% variance. The percent sales of merchandise in total operating revenue here is 79% of the estimated variance. The percent cost of merchandise sold in total operating expenses here is 74% variance (see **Table** 7).

Figure 2 presents the amount of variance for each considered variable represented by the communalities. From the aspect of PCA and Sparse PCA it can be noticed that variable Percent fuel and energy cost in total operating expenses and variable Percent finished products in total assets have significant estimated variance. When it comes to Robust PCA, variance of 82% refers to the variable Percent wage cost in total operating expenses. From the economic point of view first two variables could be used to distinguish type of three major business activities. Mainly, the amount of fuel and energy cost will differ between business activities. It is expected that production entities will have higher values of fuel and energy costs because plant, machinery and equipment will require energy to operate. Also, merchandise entities will probably have higher values of fuel and energy costs compared to other services having in mind

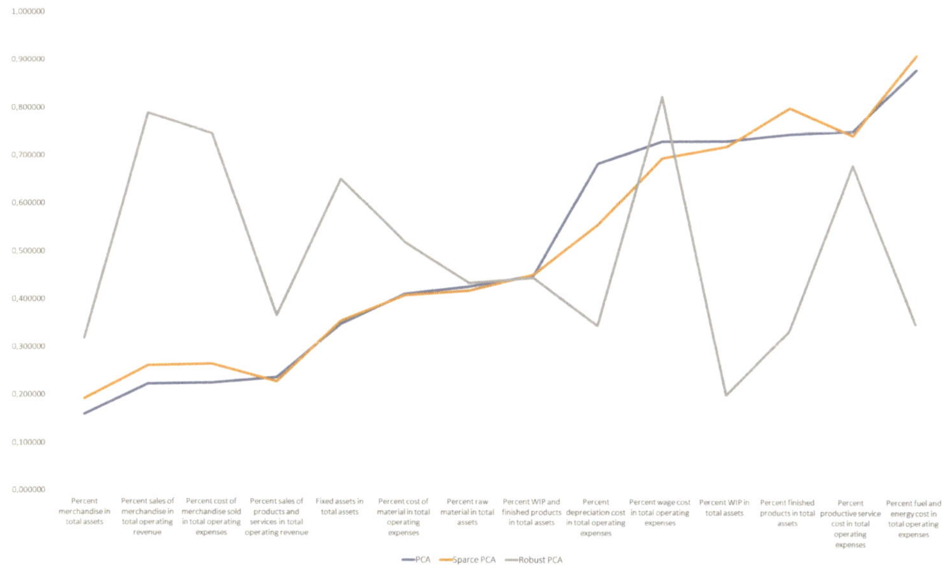

Figure 2.
Amount of variance represented by the communalities.

fuel spent for transportation of merchandise and energy needed for operation of their facilities. Second variable Percent finished products in total assets is also expected to be used for differentiation since only production entities will have this balance sheet line in their financial statements. Main surprise might be third variable Percent wage cost in total operating expenses, since most entities have very similar share of total wage costs in total operating expenses. Namely, although official state records showed that average wages differ across industries, management of companies usually plan operating expenses and their structure.

4.3 Clustering

The best approach for the PCA/Clustering combination regarding high level of Silhouette Index and Cluster Sizes are: K-means/Robust PCA and Spectral/Robust PCA. The Davies Bouldin Index implies that a smaller value gives better clustering. This produces the idea that no cluster has to be similar to another, and that object inside clusters are very uniformly distributed (see **Table 8**).

Clustering/PCA method	Cluster sizes	Silhouette index	Davies bouldin index
K-means/No PCA	(1345, 932, 733)	0.30208710358306756	1.5444364169813884
K-means/PCA	(1353, 934, 723)	0.3637346841903855	1.3405097768944103
K-means/Sparse PCA	(1356, 939, 715)	0.36307616530243575	1.3418713066940657
K-means/Robust PCA	(1209, 944, 857)	0.5193200382282146	0.7834359567299072
Agglomerative/no PCA	(1151, 935, 924)	0.27839422485839554	1.7150687814273013
Agglomerative/ PCA	(1225, 962, 823)	0.31642069773357084	1.4995739243069988

Clustering/PCA method	Cluster sizes	Silhouette index	Davies bouldin index
Agglomerative/sparse PCA	(1888, 893, 229)	0.31642069773357084	1.4995739243069988
Agglomerative/robust PCA	(1311, 878, 821)	0.4593880561940543	0.9274868826361716
Birch/no PCA	(1151, 935, 924)	0.27839422485839554	1.7150687814273013
Birch/ PCA	(1225, 962, 823)	0.31642069773357084	1.4995739243069988
Birch/sparse PCA	(1225, 962, 823)	0.31642069773357084	1.4995739243069988
Birch/robust PCA	(1317, 867, 826)	0.45631070311567473	0.9348852316431389
Gaussian mixture/no PCA	(1336, 992, 682)	0.17495781525891207	2.1078218204567496
Gaussian mixture/ PCA	(1161, 1155, 694)	0.2539355374019169	1.6227017939395394
Gaussian mixture/sparse PCA	(1161, 1155, 694)	0.2539355374019169	1.6227017939395394
Gaussian mixture/robust PCA	(1467,784, 759)	0.28455634384131373	1.1919962215015028
Spectral/no PCA	(2994, 8, 8)	0.460433642421337	0.9718901349784725
Spectral/PCA	(3001, 7, 2)	0.5399338738262545	0.6856986473871954
Spectral/sparse PCA	(3001, 7, 2)	0.5399338738262545	0.6856986473871954
Spectral/robust PCA	(1346, 920, 744)	0.5146721760042233	0.7917964357887189

Table 8.
PCA with different clustering methods.

5. Conclusion

This chapter was focused on the use of Principle component analysis in financial data science. Research has been conducted that included 3013 medium and large business entities and their financial statements from 2019 reporting period. PCA has been used in order to differentiate between the three major types of business activities - merchandising, manufacturing, and service. Therefore, 14 financial ratios have been selected by common sense and further analyzed according to their significance in dimensionality reduction. Results of clustering gave 7 new variables: 1. cost of merchandise sold in total operating expenses, and cost of material in total operating expenses; 2. fuel and energy cost in total operating expenses, and sales of product and services in total operating revenue; 3. wage costs in total operating expenses, and sales on merchandise in total operating revenue; 4. productive service cost in total operating expanses, and fixed assets in total assets; 5. depreciation cost in total operating expenses, and merchandise in total assets; 6. raw material in total assets, and WIP and finished products in total assets; 7. finished products in total assets, and WIP in total assets. These groups of variables were able to explain 86.7% of initial variability. Compared to the results of authors previously mentioned in literature review, it can be concluded that percentage is within the range of reached results. When it comes to initial communalities which estimated the variance in each variable, three financial ratios that had the highest percentage were: fuel and energy cost in total operating expenses (original PCA—88%, sparse PCA—91%); productive service cost in total operating expenses (original PCA—75%, sparse PCA—74%); and finished products in total assets (original PCA 75%, sparse PCA—80%). Although these ratios showed the best results, it has to be mentioned that there is a correlation between all of financial ratios used in analysis and therefore results would be different when ratios are used.

Acknowledgements

We would like to express our gratitude to Prof. Nemanja Stanišić, Ph.D. from the Singidunum University for supporting this research through valuable suggestions, and assignment of a research database.

Author contribution

Author Stefana Janićijević contributed to the design and implementation of the research and analysis of the results. Authors Vule Mizdraković and Maja Kljajić prepared sections of the chapter that refers to the financial data science and financial reporting: introduction, related work, research methodology and analysis of discussion and result. All authors provided critical feedback and helped shape the research, analysis, and manuscript.

Conflicts of interest

Authors declare no conflict of interest.

Nomenclature

m	number of numerical variables
n	individuals
x	vector
X	data matrix
j	number of columns
Xc	linear combinations
c	vector of constants
M	covariance matrix
λ	lagrange multiplier
U	matrix with orthonormal colums—eigenvectors
A	matrix with singular vectors
L	diagonal elements of the matrix
L^2	diagnal matrix with one square of the singular values
r	rank of the matrix
q	dimensional subspace
tr	trace of matrix

A. Appendix

See **Tables 9–11**.

Columns/factors	Factor 0	Factor 1	Factor 2	Factor 3	Factor 4	Factor 5	Factor 6
Fixed assets in total assets	0.178413	−0.326641	0.415221	−0.102354	−0.025277	−0.072754	−0.141675
Percent sales of merchandise in total operating revenue	−0.436002	0.152729	0.080029	−0.025182	0.028728	0.016652	−0.025519
Percent sales of products and services in total operating revenue	0.398117	0.022315	−0.270570	−0.046006	0.031509	0.012930	−0.028959
Percent cost of merchandise sold in total operating expenses	−0.432559	0.162995	0.080296	−0.035352	0.022542	0.026778	−0.041260
Percent cost of material in total operating expenses	0.269688	0.303749	−0.078000	−0.386050	−0.243323	−0.066637	−0.166323
Percent fuel and energy cost in total operating expenses	0.150958	−0.217356	0.283494	−0.008988	0.096350	0.822637	−0.210100
Percent wage cost in total operating expenses	0.210317	−0.224488	−0.145697	0.058149	0.719422	−0.304476	−0.022063
Percent productive service cost in total operating expenses	0.137457	−0.048360	−0.397081	0.585499	−0.374661	0.172006	0.245674
Percent depreciation cost in total operating expenses	0.095815	−0.269993	0.484683	0.000289	−0.400868	−0.359862	0.275725
Percent raw material in total assets	0.190490	0.245296	−0.165694	−0.526444	−0.137683	0.071830	0.032101
Percent WIP in total assets	0.158335	0.359273	0.200936	0.383609	−0.059087	−0.175372	−0.596528
Percent finished products in total assets	0.174390	0.375283	0.278149	0.015943	0.252221	0.151402	0.640266
Percent WIP and finished products in total assets	0.214830	0.474174	0.309621	0.255892	0.126328	−0.013709	0.034896
Percent merchandise in total assets	−0.355975	0.151166	−0.014079	−0.013559	0.088827	0.039431	0.005508

Table 9.
PCA component matrix.

Columns/factors	Factor 0	Factor 1	Factor 2	Factor 3	Factor 4	Factor 5	Factor 6
Fixed assets in total assets	0.000000	−0.020910	−0.504987	0.000000	−0.254639	−0.185601	−0.002365
Percent sales of merchandise in total operating revenue	0.435472	0.000000	0.246576	−0.085566	0.000000	0.052803	0.000000
Percent sales of products and services in total operating revenue	−0.433624	0.008108	0.000000	0.199284	0.000000	0.000000	0.000000
Percent cost of merchandise sold in total operating expenses	0.438993	0.000000	0.254341	−0.067395	0.000000	0.044065	0.000000
Percent cost of material in total operating expenses	−0.027509	0.085834	0.000000	0.630267	0.000000	0.045436	−0.019978
Percent fuel and energy cost in total operating expenses	0.000000	0.000000	0.000000	0.000000	0.000000	−0.954607	0.000000
Percent wage cost in total operating expenses	−0.453726	0.000000	0.000000	−0.325539	−0.594326	0.173439	0.000000
Percent productive service cost in total operating expenses	−0.333679	0.000000	0.000000	−0.222344	0.762847	0.000000	0.000000
Percent depreciation cost in total operating expenses	0.108694	0.000000	−0.726249	0.000000	0.000000	0.128099	0.000000
Percent raw material in total assets	0.000000	−0.007293	0.083372	0.624407	−0.000201	0.000000	0.143399
Percent WIP in total assets	0.000000	0.460374	0.000000	0.000000	0.000000	0.000000	−0.712393
Percent finished products in total assets	0.000000	0.573218	0.000000	0.000000	0.000000	0.000000	0.686680
Percent WIP and finished products in total assets	0.000000	0.671977	0.000000	0.000000	0.000000	0.000000	0.000000
Percent merchandise in total assets	0.315974	0.000000	0.291738	−0.076742	0.000000	0.031515	0.000000

Table 10.
Sparse PCA component matrix.

Columns/factors	Factor 0	Factor 1	Factor 2	Factor 3	Factor 4	Factor 5	Factor 6
Fixed assets in total assets	−0.173467	0.275000	0.507855	−0.499215	0.156525	−0.088965	0.078114
Percent sales of merchandise in total operating revenue	0.525938	−0.128407	0.122190	−0.093645	−0.109748	−0.078270	0.673835
Percent sales of products and services in total operating revenue	−0.444479	−0.119035	−0.307493	−0.162452	−0.114015	−0.009416	−0.142249
Percent cost of merchandise sold in total operating expenses	0.510523	−0.126489	0.118665	−0.111772	−0.087492	0.088045	−0.653626
Percent cost of material in total operating expenses	−0.204519	−0.558377	0.030438	−0.282440	−0.265091	0.084341	0.087885
Percent fuel and energy cost in total operating expenses	−0.119620	0.103472	0.453140	0.056594	−0.245465	0.200560	−0.125820
Percent wage cost in total operating expenses	−0.204794	0.179552	0.159958	0.368300	−0.273464	−0.715834	−0.010912
Percent productive service cost in total operating expenses	−0.131802	0.032733	0.012123	0.533088	−0.234838	0.537258	0.183658
Percent depreciation cost in total operating expenses	−0.098392	0.135731	0.478665	0.038015	−0.137017	0.259630	−0.023300
Percent raw material in total assets	−0.120240	−0.430495	0.139923	−0.141176	−0.424059	−0.117921	0.026698
Percent WIP in total assets	−0.048543	−0.251299	0.164302	0.160983	0.282957	−0.044589	−0.000473
Percent finished products in total assets	−0.062609	−0.324119	0.211913	0.207631	0.364950	−0.057519	0.022279
Percent WIP and finished products in total assets	−0.071814	−0.371772	0.243069	0.238158	0.418607	−0.065970	−0.072968
Percent merchandise in total assets	0.289968	−0.108389	0.082710	0.230665	−0.308640	−0.196863	−0.167039

Table 11.
Robust PCA component matrix.

DOI: http://dx.doi.org/10.5772/intechopen.102928

Author details

Stefana Janićijević*, Vule Mizdraković and Maja Kljajić
Singidunum University, Belgrade, Republic of Serbia

*Address all correspondence to: sjanicijevic@singidunum.ac.rs

References

[1] Mrvaljevic M, Dobricanin S, Djuricanin J. Finansijski izveštaji u funkciji menadžment odlučivanja. Ekonomski signali. 2014;**9**(2):85-103

[2] Djukic T, Pavlovic M. Kvalitet finansijskog izveštavanja u republici srbiji. Ekonomske teme. 2014;**52**(1): 101-116

[3] Law on Accounting ("Off. Herald of RS", Nos. 62/2013, 30/2018 and 73/2019 - other law). 2021

[4] Bhunia A, Mukhuti SS, Roy SG. Financial performance analysis—A case study. Current Research Journal of Social Sciences. 2011;**3**(3):269-275

[5] Abraham A. A model of financial performance analysis adapted for nonprofit organisations. In: AFAANZ 2004 Conference Proceedings. Melbourne, Austalia: Accounting & Finance Association of Australia and New Zealand (afaanz) Limited; 2004. pp. 1-18

[6] Bhargava P. Financial analysis of information and technology industry of India (A Case Study of Wipro Ltd and Infosys Ltd). Journal of Accounting, Finance and Auditing Studies. Yalova, Turkey: Istanbul Business Academy. 2017; **3**(3):1-13

[7] Schönbohm A. Performance Measurement and Management with Financial Ratios: The Basf se Case. Technical report, Working Paper. London: IEEE; 2013

[8] Taylor SL. Analysing financial statements: How many variables should we look at? JASSA. 1986;**1**(1):19-21

[9] Karamizadeh S, Abdullah SM, Manaf AA, Zamani M, Hooman A. An overview of principal component analysis. Journal of Signal and Information Processing. 2013; **4**(3B):173

[10] Abbas AH, Arab A, Harbi J. Image compression using principal component analysis. Mustansiriyah Journal of Science. 2018;**29**(2):01854

[11] Polyak BT, Khlebnikov MV. Principle component analysis: Robust versions. Automation and Remote Control. 2017;**78**(3):490-506

[12] Dunteman GH. Principal Components Analysis. Thousand Oaks, California, United States: Sage; 1989. p. 69

[13] Jacod J, Lejay A, Talay D. Estimation of the brownian dimension of a continuous itô process. Bernoulli. 2008; **14**(2):469-498

[14] Rummel RJ. Applied Factor Analysis. Evanston, Illinois, United States: Northwestern University Press; 1988

[15] Demmel JW. Applied Numerical Linear Algebra, SIAM. Philadelphia, Pennsylvania, United States: Society for Industrial and Applied Mathematics; 1997

[16] Abdi H, Williams LJ. Principal component analysis. Wiley Interdisciplinary Reviews: Computational Statistics. 2010;**2**(4): 433-459

[17] Tibshirani R. Regression shrinkage and selection via the lasso. Journal of the Royal Statistical Society: Series B (Methodological). 1996;**58**(1):267-288

[18] Hastie T, Tibshirani R, Wainwright M. The lasso for linear models. Statistical Learning with

Sparsity: The LASSO and Generalization. 2015;**7–28**

[19] Xu H, Caramanis C, Sanghavi S. Robust PCA Via Outlier Pursuit. arXiv preprint arXiv:1010.4237. 2010

[20] Candès EJ, Li X, Ma Y, Wright J. Robust principal component analysis? Journal of the ACM (JACM). 2011;**58**(3): 1-37

[21] Wright J, Ganesh A, Rao S, Peng Y, Ma Y. Robust principal component analysis: Exact recovery of corrupted low-rank matrices via convex optimization. Advances in Neural Information Processing Systems. 2009; **58**:289-298

[22] Pearson K. LIII. On lines and planes of closest fit to systems of points in space. The London, Edinburgh, and Dublin Philosophical Magazine and Journal of Science. 1901;**2**(11): 559-572

[23] Hotelling H. Analysis of a complex of statistical variables into principal components. Journal of Educational Psychology. 1933;**24**(6):417

[24] Litterman R, Scheinkman J. Common factors affecting bond returns. Journal of Fixed Income. 1991;**1**(1):54-61

[25] Stock JH, Watson MW. Forecasting inflation. Journal of Monetary Economics. 1999;**44**(2):293-335

[26] Egloff D, Leippold M, Wu L. The term structure of variance swap rates and optimal variance swap investments. Journal of Financial and Quantitative Analysis. 2010;**45**(5):1279-1310

[27] Baker M, Wurgler J. Investor sentiment and the cross-section of stock returns. The Journal of Finance. 2006; **61**(4):1645-1680

[28] Baker SR, Bloom N, Davis SJ. Measuring economic policy uncertainty. The Quarterly Journal of Economics. 2016;**131**(4):1593-1636

[29] Anderson DL, Weeks WF. A theoretical analysis of sea-ice strength. Eos, Transactions American Geophysical Union. 1958;**39**(4):632-640

[30] Anderson OL. A simplified method for calculating the debye temperature from elastic constants. Journal of Physics and Chemistry of Solids. 1963;**24**(7): 909-917

[31] Waternaux CM. Asymptotic distribution of the sample roots for a nonnormal population. Biometrika. 1976;**63**(3):639-645

[32] Jolliffe D. Whose education matters in the determination of household income? Evidence from a developing country. Economic Development and Cultural Change. 2002;**50**(2):287-312

[33] Tyler WG. Growth and export expansion in developing countries: Some empirical evidence. Journal of Development Economics. 1981;**9**(1): 121-130

[34] Brillinger DR. Time Series: Data Analysis and Theory. Philadelphia, Pennsylvania, United States: SIAM. Society for Industrial and Applied Mathematics; 2001

[35] Jacod J, Podolskij M. A test for the rank of the volatility process: The random perturbation approach. The Annals of Statistics. 2013;**41**(5): 2391-2427

[36] Mykland PA, Zhang L. Inference for continuous semimartingales observed at high frequency. Econometrica. 2009; **77**(5):1403-1445

[37] Li J, Todorov V, Tauchen G. Volatility occupation times. The Annals of Statistics. 2013;**41**(4):1865-1891

[38] Saha S, Moorthi S, Pan H-L, Wu X, Wang J, Nadiga S, et al. The ncep climate forecast system reanalysis. Bulletin of the American Meteorological Society. 2010;**91**(8):1015-1058

[39] Tao M, Wang Y, Yao Q, Zou J. Large volatility matrix inference via combining low-frequency and high-frequency approaches. Journal of the American Statistical Association. 2011;**106**(495): 1025-1040

[40] Jacod J, Rosenbaum M. Quarticity and other functionals of volatility: Efficient estimation. The Annals of Statistics. 2013;**41**(3):1462-1484

[41] Heinrich C, Podolskij M. On Spectral Distribution of High Dimensional Covariation Matrices. arXiv preprint arXiv:1410.6764. 2014

[42] Kalnina I, Xiu D. Nonparametric estimation of the leverage effect: A trade-off between robustness and efficiency. Journal of the American Statistical Association. 2017;**112**(517): 384-396

[43] Jara EG, Ebrero AC, Zapata RE. Effect of international financial reporting standards on financial information quality. Journal of Financial Reporting and Accounting. 2011;**9**(2):176-196

[44] Lantto A-M, Sahlström P. Impact of international financial reporting standard adoption on key financial ratios. Accounting & Finance. 2009; **49**(2):341-361

[45] Triandafil C, Brezeanu P, Badea L. Impactul macroeconomic asupra profotabilitagii sectorului corporativ: analiza la nivelul companiilor listate la bursa de valori bucuresti. Economie teoretica si aplicata. 2010;**17**(10):551

[46] Tudor E. Metode de recunoastere a formelor in analiza economico-financiara. Bucharest: Academy of Economic Studies; 2009

[47] Armeanu D, Negru A. Aplicarea analiza componentelor principale in managementul portofoliului de investitii. Internal Auditing & Risk Management. 2011;**6**(3):01865

[48] Robu IB, Istrate C. The analysis of the principal components of the financial reporting in the case of Romanian listed companies. Procedia Economics and Finance. 2015;**20**:553-561

[49] Mizdrakovic V, Stanic N, Mitic V, Obradovic A, Kljajic M, Obradovic M, et al. Empirical data on financial and audit reports of Serbian business entities. In: FINIZ 2020-People in the Focus of Process Automation. Belgrade, Serbia: Singidunum University International Scientific Conference; 2020. pp. 193-198

Chapter 8

Determining an Adequate Number of Principal Components

Stanley L. Sclove

Abstract

The problem of choosing the number of PCs to retain is analyzed in the context of *model selection,* using so-called *model selection criteria* (MSCs). For a prespecified set of models, indexed by $k = 1, 2, \ldots, K$, these model selection criteria (MSCs) take the form $\text{MSC}_k = nLL_k + a_n m_k$, where, for model k, LL_k is the maximum log likelihood, m_k is the number of independent parameters, and the constant a_n is $a_n =$ $\ln n$ for BIC and $a_n - 2$ for AIC. The maximum log likelihood LL_k is achieved by using the maximum likelihood estimates (MLEs) of the parameters. In Gaussian models, LL_k involves the logarithm of the mean squared error (MSE). The main contribution of this chapter is to show how to best use BIC to choose the number of PCs, and to compare these results to *ad hoc* procedures that have been used. Findings include the following. These are stated as they apply to the eigenvalues of the correlation matrix, which are between 0 and p and have an average of 1. For considering an additional PC_{k+1}, with AIC, inclusion of the additional PC_{k+1} is justified if the corresponding eigenvalue λ_{k+1} is greater than $\exp(-2/n)$. For BIC, the inclusion of an additional PC_{k+1} is justified if $\lambda_{k+1} > n^{1/n}$, which tends to 1 for large n. Therefore, this is in approximate agreement with the average eigenvalue rule for correlation matrices, stating that one should retain dimensions with eigenvalues larger than 1.

Keywords: reduction of dimensionality, principal components, model selection criteria, information criteria, AIC, BIC

1. Introduction and background

1.1 Introduction

Sometimes, researchers know how many principal components (PCs) they need. For example, to construct an optimal scatterplot, the scores of the sample on the first two principal components will be used to obtain an optimal plot. For an optimal three-dimensional scatterplot, the scores on the first three principal components will be used. In many applications, however, the researchers will question how many princi pal components they need. This chapter discusses the application of various methods to the problem of reduction of dimensionality, in the sense of choosing an adequate number of principal components to retain to represent a dataset. The methods discussed include *ad hoc* methods, likelihood-based methods, and model selection criteria (MSCs), especially Akaike's information criterion (AIC) and Bayesian information criterion (BIC). This chapter applies the concepts of [1, 2] to this particular problem.

1.2 Background

To begin the discussion here, we first give a short review of some general background on the relevant portions of multivariate statistical analysis, which may be obtained from textbooks such as [3] or [4].

1.3 Sample quantities

Let $\mathbf{x}_1, \mathbf{x}_2, \ \dots, \ \mathbf{x}_n$ denote a sample of n p-dimensional random vectors

$$\mathbf{x}_i = \left(x_{1i}, x_{2i}, \dots . \ x_{pi}\right)', \qquad i = 1, 2, \dots, n. \tag{1}$$

Here, the transpose ($'$) means that the vectors are being considered as column vectors. The sample *mean vector* is

$$\overline{\mathbf{x}} = \sum_{i=1}^{n} \mathbf{x}_i / n. \tag{2}$$

The $p \times p$ sample covariance matrix is denoted by

$$\mathbf{S} = \sum_{i=1}^{n} (\mathbf{x}_i - \overline{\mathbf{x}})(\mathbf{x}_i - \overline{\mathbf{x}})' / (n-1). \tag{3}$$

1.4 Population quantitites

The sample covariance matrix **S** estimates the true covariance matrix Σ of the random variables

$$X_1, X_2, \dots, X_p.$$

The true covariance matrix is

$$\sum = [\sigma_{u,v}]_{u,v=1,2,\dots,p}, \tag{4}$$

where

$$\sigma_{uv} = C[X_u, X_v], \tag{5}$$

the covariance of X_u and X_v, for $u \neq v, \ \ u, v = 1, 2, \dots, p$. For $u = v$, we have $C[X_v, X_v] = V[X_v]$, the variance of X_v.

1.5 Principal components

The principal components of Σ are defined as *uncorrelated linear combinations of maximal variance.* Let us elaborate on this brief definition. First, a linear combination, say LC, of the p variables can be expressed as the vector product $\mathbf{a}'\mathbf{x}$ of two vectors $\mathbf{a}$ and $\mathbf{x}$, that is,

$$LC = \mathrm{a}'\mathbf{x} = a_1x_1 + a_2x_2 + \cdots + a_px_p. \quad (6)$$

Here, the vector $\mathbf{a}$ is a vector of scalars $a_1, a_2, \ldots, a_p$:

$$\mathrm{a}' = \begin{pmatrix} a_1 & a_2 & \ldots & a_p \end{pmatrix}. \quad (7)$$

These a_j are the coefficients in the linear combination. Such linear combinations are called *variates.* Principal components are also called *latent variables.*

The variance $\mathcal{V}$ of a linear combination LC is

$$\mathcal{V}[LC] = \mathcal{V}[\mathrm{a}'\mathrm{X}] = \mathrm{a}'\Sigma\mathrm{a}. \quad (8)$$

This is estimated as $\mathbf{a}'\mathbf{Sa}$. This is to be maximized over $\mathbf{a}$. The derivative with respect to the vector $\mathbf{a}$ is

$$\partial\mathrm{a}'\mathrm{Sa}/\partial\mathrm{a} = \mathrm{Sa}. \quad (9)$$

The solution is not unique: If $\mathbf{a}$ is a solution to this set of equations, so is $c\mathbf{a}$, where c is any scalar constant. Therefore, a constraint is required to obtain a meaningful solution. A reasonable such constraint is the condition $\mathbf{a}'\mathbf{a} = 1$, that is, the squared length of the vector $\mathbf{a}$ equals 1. This is of course equivalent to the length of $\mathbf{a}$, the quantity $\sqrt{\mathbf{a}'\mathbf{a}}$, being equal to 1.

A function incorporating the constraint, the *Lagrangian function,* is

$$L(\mathrm{S}, \mathrm{a}; \lambda) = \mathrm{a}'\mathrm{Sa} + \lambda(1 - \mathrm{a}'\mathrm{a}). \quad (10)$$

The partial derivatives of the function L with respect to a and λ are

$$\partial L/\partial\mathrm{a} = 2\mathrm{Sa} - 2\lambda\mathrm{a} \quad (11)$$

and

$$\partial L/\partial\lambda = \partial\lambda(1 - \mathrm{a}'\mathrm{a})/\partial\lambda = 1 - \mathrm{a}'\mathrm{a}. \quad (12)$$

Setting these partial derivatives equal to zero gives the simultaneous linear equations

$$\mathrm{Sa} = \lambda\mathrm{a}, \quad (13)$$

and the equation

$$\mathrm{a}'\mathrm{a} = 1. \quad (14)$$

The simultaneous linear equations can be written as

$$\mathrm{Sa} - \lambda \mathrm{a} = 0, \tag{15}$$

where **0** is the zero vector, the vector whose elements are all zeroes. Factoring out **a** on the right, we obtain

$$(\mathrm{S} - \lambda \mathrm{I})\mathrm{a} = 0. \tag{16}$$

For nontrivial solutions, the determinant of the coefficient matrix $\mathbf{S} - \lambda\mathbf{I}$ must be zero, that is, we must have $\det(\mathbf{S} - \lambda\mathbf{I}) = 0$. This condition is a polynomial equation of degree p in λ. Denote the p roots by $\lambda_1 \geq \lambda_2 \geq \cdots \geq \lambda_p$. These roots are the *eigenvalues* (also called *latent values*). Their sum is the trace of **S**; their product is the determinant of **S**.

The corresponding Eigen equations are

$$\mathrm{Sa}_j = \lambda_j \mathrm{a}_j, \quad j = 1, 2, \ldots, p. \tag{17}$$

1.5.1 Values of PCs in terms of Xs

The jth principal component (PC), C_j, is the linear combination of the form

$$C_j = \mathrm{a}'_j \mathrm{x} = a_{1j}x_1 + a_{2j}x_2 + \cdots + a_{pj}x_p, \tag{18}$$

where $\mathbf{a}'_j = (a_{1j}, a_{2j}, \ldots, a_{pj})$. That is to say, for $j = 1, 2, \ldots, p$, the value of the jth PC for Individual i is $c_{ji} = \mathbf{a}'_j \mathbf{x}_i$, $i = 1, 2, \ldots, n$..

The equation for the jth PC in terms of the vector $\mathbf{x} = (x_1 x_2 \ldots x_p)'$ is $c_j = \mathbf{a}'_j \mathbf{x}$, $j = 1, 2, \ldots, p$. Let **c** be the p-vector of values of the p PCs. Then, $\mathbf{c} = \mathbf{A}'\mathbf{x}$, where $\mathbf{A} = [\mathbf{a}_1 \ \mathbf{a}_2 \ \ldots \ \mathbf{a}_p]$ is the $p \times p$ matrix whose columns are the eigenvectors.

1.5.2 Values of Xs in terms of PCs

The inverse relation is

$$\mathrm{x} = \mathrm{A}'^{-1}\mathrm{c} = \mathrm{Bc}, \tag{19}$$

where

$$\mathrm{B} = \mathrm{A}'^{-1}, \tag{20}$$

where **B** is the matrix of *loadings* of the X_v on the PCs C_j. Actually, **A** is an orthonormal matrix (meaning that its columns are of length one and are pairwise orthogonal), so $\mathbf{A}^{-1} = \mathbf{A}'$. Thus, $\mathbf{B} = \mathbf{A}$. Therefore,

$$\mathrm{x} = \mathrm{A}'^{-1}\mathrm{c} = \mathrm{Ac}. \tag{21}$$

Letting $\mathbf{a}^{(v)'}$ be the vth row of the matrix **A**, that is,

$$\mathbf{a}^{(v)'} = (a_{v1}, a_{v2}, \ldots, a_{vp}), \tag{22}$$

we have, for $v = 1, 2, \dots, p$, the representation of each variable X_v in terms of the variables $C_1, C_2, \dots, C_p$ that are the principal components,

$$X_v = a_{v1}C_1 + a_{v2}C_2 + \cdots + a_{vp}C_p. \tag{23}$$

In terms of the first k PCs, this is

$$X_v = a_{v1}C_1 + a_{v2}C_2 + \cdots + a_{vk}C_k + \varepsilon_v, \quad (*) \tag{24}$$

where the error ε_v is

$$\varepsilon_v = a_{v\,k+1}C_{k+1} + a_{v\,k+2}C_{k+2} + \cdots + a_{vp}C_p. \tag{25}$$

The covariance matrix can be represented in terms of its *principal idempotents* $\mathbf{a}_j\mathbf{a}'_j$ as

$$S = \sum_{j=1}^{p} \lambda_j a_j a'_j. \tag{26}$$

It follows as a result of this representation that the best approximation of rank k to S is the eigenvalue weighted sum of the first k principal idempotents,

$$S^{(k)} = \sum_{j=1}^{k} \lambda_j a_j a'_j. \tag{27}$$

The weights are all non-negative, recalling that, for a symmetric matrix, such as a covariance matrix, the eigenvalues are non-negative.

2. Some *ad hoc* arithmetic procedures for determining an appropriate number of PCs

2.1 Procedure based on the average of the eigenvalues

The mean $\bar{\lambda}$ of the eigenvalues is the sum over the number

$$\bar{\lambda} = \sum_{j=1}^{p} \lambda_j / p. \tag{28}$$

The sum of the eigenvalues turns out to be equal to the *trace* of the covariance matrix; therefore, the mean eigenvalue is equal to the trace divided by p.

One procedure for deciding on the number of PCs to retain is to retain those for which the eigenvalues are greater than average, that is, greater than $\bar{\lambda}$. When working in terms of the correlation matrix, this average value is 1. To see this, recall that the correlation matrix is a special case of the covariance matrix, namely, the correlation matrix is the covariance matrix of the standardized variables. It is often preferable to work in terms of the correlation matrix rather than the covariance matrix, to control the effects of different units of measurement and different variances. If a variable has high variance relative to the other variables, the PC will be pulled in the direction of the variable with large variance.

When **S** is taken to be the sample *correlation* matrix, the trace of the matrix is simply p, and therefore, the mean $\overline{\lambda}$ of the eigenvalues is 1.

2.2 An *ad hoc* arithmetic procedure based on retaining a prescribed proportion of the total variance

Another *ad hoc* procedure is to retain a number of PCs sufficient to account for a prescribed proportion, say, 90% of the total variance, that total variance being trace $\mathbf{S} = \sum_{j=1}^{p} \lambda_j$. The Figure 90% is of course somewhat arbitrary, so it might be good to have some somewhat more objective criteria based on the pattern of the eigenvalues.

2.3 Procedure based on the decrease of the eigenvalues

Another procedure—a graphical procedure—is to plot $\lambda_1, \lambda_2, \dots, \lambda_p$ against $1, 2, \dots, p$. The λs are in decreasing order, so one then looks for a dropoff—an elbow—in the curve and retains a number of PCs corresponding to the point before the leveling off of the curve, if it does indeed take an elbow shape. Such a plot, of the eigenvalues versus $1, 2, \dots, p$, is called a *scree* plot, "scree" being the debris at the foot of a glacier (or, more generally, a collection of broken rock fragments at the base of crags, mountain cliffs, volcanoes, or valley shoulders).

3. Model selection criteria AIC and BIC for the number of PCs

Let us now delve a bit further into mathematical statistics and consider some more objective, numerical criteria, in particular, the information criteria AIC and BIC. Let us see what a Gaussian model would imply about AIC and BIC. The maximum log likelihood for the model (*) approximating the p variables in terms of k PCs is $(2\pi)^{-np/2} |\hat{\Sigma}_k|^{-n/2} C(n,p,k)$, where $C(n,p,k)$ is a constant depending upon the sample size, n, the number of variables, p, and k, the Model k being considered, $k = 1, 2, \dots, K$, and $|\Sigma_k|$ denotes the determinant of the residual covarance matrix Σ_k.

The determinant of the covariance matrix is the product of the eigenvalues,

$$|\Sigma| = \prod_{j=1}^{p} \lambda_j. \tag{29}$$

For a model based on the first k PCs, the determinant of the residual covariance matrix is the product of the remaining, smaller eigenvalues, $\Pi_{j=k+1}^{p} \lambda_j$.

The model selection criterion AIC—Akaike's information criterion [5–7]—is based on an estimate of the logarithm of the cross-entropy of the K proposed models with a null model. That is, for alternative models indexed by $k = 1, 2, \dots, K$, AIC_k is an estimate of the log cross-entropy of the proposed Model k with the null model. The cross-entropy of the distribution with the probability density function $q(\mathbf{x})$ relative to a distribution with the probability density function $p(\mathbf{x})$ is defined as $H(p,q) = -\mathcal{E}_p[\ln q(\mathbf{X})] = -\int \ln q(\mathbf{x}) p(\mathbf{x}) \, d\mathbf{x}$.

The Bayesian information criterion (BIC) [8] is based on a large-sample estimate of the posterior probability pp_k of Model k, $k = 1, 2, \dots, K$. More precisely, BIC_k is an approximation to $-2 \ln pp_k$.

Formulated in this way, these model selection criteria (MSCs) are, thus, smaller-is-better criteria and take the form

$$MSC_k = -2 \ln \max L_k + a_n m_k, \qquad k = 1, 2, \ldots, K, \tag{30}$$

where L_k is the likelihood for Model k, $a_n = \ln n$ for BIC_k, $a_n = 2$ (not depending upon n) for AIC_k, and m_k is the number of independent parameters in Model k. The first term is a lack-of-fit (LOF) term, and the second term is a penalty term based on the number of parameters used. With AIC, the penalty is two units per parameter; with BIC, the penalty is $\ln n$ units per parameter. For $n \geq 8$, $\ln n$ exceeds 2: for sample sizes greater than 7, the penalty per parameter with BIC exceeds that for AIC. Therefore, relative to AIC, BIC tends to favor more parsimonious models—models with a smaller number of parameters.

Note that

$$pp_k \approx C \exp(-\text{BIC}_k/2), \tag{31}$$

where C is a constant. Thus, BIC values can be converted to values on a scale of 0–1. This is done by exponentiating $-\text{BIC}_k/2$, summing the values, and dividing by the sum. That is,

$$pp_k \approx \exp(-\text{BIC}_k/2) / \sum_{j=1}^{K} \exp\left(-\text{BIC}_j/2\right). \tag{32}$$

To relate the maximum likehood to the eigenvalues, note that for the PC model,

$$-2 \ln \max L_k = n \ln \Pi^{p}_{j=k+1} \lambda_k = n \sum_{j=k+1}^{p} \ln \lambda_k. \tag{33}$$

The model selection criteria can be written as

$$\text{MSC}_k = \text{Deviance}_k + \text{Penalty}_k, \tag{34}$$

where $\text{Deviance}_k = n \ln \max L_k$ is a measure of lack of fit and $\text{Penalty}_k = a_n m_k$. Inclusion of an additional PC is justified if the criterion value decreases, that is, if $\text{MSC}_{k+1} < \text{MSC}_k$. For PCs, this is

$$n \sum_{j=k+2}^{p} \ln \lambda_j + (k+1) a_n < n \sum_{j=k+1}^{p} \ln \lambda_j + k a_n. \tag{35}$$

This is

$$a_n < n \ln \lambda_{k+1} = \ln\left(\lambda^n_{k+1}\right), \tag{36}$$

or

$$\exp[a_n] < \lambda^n_{k+1}, \tag{37}$$

or

$$\lambda_{k+1} > \exp\left[a_n/n\right] \tag{38}$$

or

$$\lambda_{k+1} > \exp\left[-a_n/n\right]. \tag{39}$$

Thus, for AIC, the inclusion of the additional PC_{k+1} is justified if λ_{k+1} is greater than $\exp(-2/n)$.

For BIC, the inclusion of an additional PC_{k+1} is justified if

$$\lambda_{k+1} > \exp(\ln n/n) = [\exp(\ln n)]^{1/n} = n^{1/n}. \tag{40}$$

The quantity $n^{1/n}$ tends to 1 for large n. Therefore, this procedure is in approximate agreement with the average eigenvalue rule for correlation matrices, stating that one should retain dimensions with eigenvalues larger than 1.

4. Examples

4.1 An artifical example

The synthesis/analysis paradigm can be useful for understanding a problem. This means synthesizing (simulating) a dataset, so that you know the model and parameter values, and then applying your analysis method to see how well it performs. In the present context, it is interesting to simulate a dataset of measurements of rectangles, with variables length (L) and width (W) and also some functions of those such as perimeter = 2 L + 2 W and difference = L–W. In one synthesis, we took L to be Normal with a mean of 10 and a variance of 1, W was Normal with a mean of 10 and a variance of 1, PERI = 2 L + 2 W plus N(0,1) error, and DIFF = L–W plus N(0,1) error. The eigensystem was computed, and as expected, it is noted that there are two large eigenvalues, with subsequent ones dropping off a lot in value and being close to zero. The eigenvalues of the correlation matrix were 1.91, 1.83, 0.21, and 0.05.

4.2 A real example

Next, we consider the principal component analysis of a sample from the Los Angeles (LA) Heart Study. This was a long-term study, 1947–1972. It was a study among Civil Servants of Los Angeles county. LA civil servants, 2252, randomly selected, ages 21–70, received a battery of examinations for "routine" cardiovascular disease (CVD) risk factors.

The variables include age, systolic blood pressure (SYS), diastolic blood pressure (DIAS), weight (WT), height (HT), and coronary incident, a binary variable indicating whether the individual had a coronary incident during the course of the study. Blood pressure is reported as a bivariate variable, (SYS, DIAS). SYS is the pressure when the heart pumps, and DIAS is the pressure when the heart relaxes.

In the textbook [9], data for a sample of $n = 100$ men were studied. (Data on the same variables for another sample of 100 men are also given in [9]. Results can be compared and contrasted between the two samples.) Although, of course, the emphasis in the Heart Study was on explaining and predicting the coronary incident variable, here, we focus on the first five variables, their representation in terms of a smaller number of PCs, and the interpretations of the PCs. we did the PC analysis; it was not in the LA Heart Study or the textbook.

We used Minitab statistical software for the analysis. Aspects of the analysis are shown as follows.

The lower-triangular portion of the correlation matrix for the five variables is shown in **Table 1**. The highest correlation is 0.835, between SYS and DIAS. The next highest correlation, 0.426, is between HT and WT.

4.3 Principal component analysis in the example

Note that an eigenvector can be multiplied by −1, changing the signs of all its elements. In the following, this is done with PC1 so that SYS and DIAS have positive loadings. Our interpretations, related to the scientific/medical context of the study, are BPtotal, SIZE, AGE, OVERWT, and BPdiff and are written below the eigenvectors. The interpretations are based on which loadings are large and which are small, that is, on the relative sizes of the loadings. Taking 0.6 as a cutoff point, in PC1, SYS and DIAS have loadings above this, while the other variables have loadings less than this (in fact, less than 0.4), so PC1 can be interpreted as an index of total BP. In PC2, the variables WT and HT have large loadings with the same sign, so PC2 can be interpreted as SIZE (**Tables 2** and **3**).

	AGE	SYS	DIAS	WT	
SYS	0.342				
DIAS	0.354	0.835			<= NOTE highest r of 0.835 is btw SYS and DIAS
WT	−0.009	0.261	0.308		
HT	−0.332	−0.088	−0.099	0.426	<= NOTE next highest r of 0.426 is btw HT and WT

Correlations: AGE, SYS, DIAS, WT, HT.
Cell Contents: Pearson correlation.

Table 1.
Correlation matrix of five variables—LA heart data.

Eigenanalysis of the correlation matrix					
Eigenvalue	2.1894	1.5382	0.6617	0.4485	0.1621
Proportion	0.438	0.308	0.132	0.090	0.032
Cumulative	0.438	0.746	0.878	0.968	1.000
Variable	PC1	PC2	PC3	PC4	PC5
AGE	−0.394	−0.365	0.800	−0.269	0.005
SYS	−0.615	0.050	−0.342	−0.174	0.687
DIAS	−0.624	0.063	−0.291	−0.049	−0.721
WT	−0.252	0.616	0.373	0.642	0.078
HT	0.117	0.694	0.141	0.695	0.051

Principal component analysis: AGE, SYS, DIAS, WT, HT.

Table 2.
PCs of heart data.

Variable	PC1	PC2	PC3	PC4	PC5
AGE	0.394	−0.365	0.800	−0.269	0.005
SYS	0.615	0.050	−0.342	−0.174	0.687
DIAS	0.624	0.063	−0.291	−0.049	−0.721
WT	0.252	0.616	0.373	0.642	0.078
HT	−0.117	0.694	0.141	−0.695	−0.051
Interpretations (edited in to the computer output):					
	BPtotal	SIZE	AGEindex	OVERWT	BPdiff

Table 3.
PC1 is multiplied by −1.

As above, denote the eigensystem in terms of the eigenpairs

$$(\lambda_v,\ \boldsymbol{a}_v),\quad v = 1, 2, \ldots, p. \tag{41}$$

Then, the eigensystem equations are

$$\mathbf{S}\ \boldsymbol{a}_v = \lambda_v\ \boldsymbol{a}_v,\quad v = 1, 2, \ldots, p. \tag{42}$$

Here, **S** is taken to be the correlation matrix. Let $\mathbf{1}_v' = (0\ \ 0\cdots\ \ 1\cdots\ \ 0\cdots\ )$, the vector with 1 in the vth position and zeroes elsewhere. The covariance between a variable X_v and a PC C_u is $\mathcal{C}[X_v, C_u] = \mathcal{C}\left[\mathbf{1}_v'\boldsymbol{X},\ \boldsymbol{a}_u'\boldsymbol{X}\right] = \mathbf{1}'\Sigma\ \ a_u = \mathbf{1}_v'\lambda_u\,\boldsymbol{a}_u = \lambda_u a_{uv}$, where a_{uv} is the vth element of the vector a_u. The coefficient of correlation is $\mathrm{Corr}[X_v, C_u] = \mathcal{C}[X_v, C_u]/\mathrm{SD}\,[X_v]\ \mathrm{SD}\ [C_u] = \lambda_u a_{uv}/\sigma_v\sqrt{\lambda_u} = \sqrt{\lambda_u}a_{uv}/\sigma_v$. When the covariance matrix used is the correlation matrix, each standard deviation $\sigma_v = 1$, and therefore, this correlation is $\sqrt{\lambda_u}\ a_{uv}$. A correlation of size greater than 0.6 corresponds to more than $0.6^2 \times 100\% = 36\%$ of variance explained. The variable X_v has a correlation higher than 0.6 with the component C_u if its loading in C_u, the value a_{uv}, is greater than $0.6 / \sqrt{\lambda_u}$. These values are appended to **Table 4**. Loadings larger than

Variable	PC1	PC2	PC3	PC4	PC5
AGE	0.394	−0.365	**0.800**	−0.269	0.005
SYS	**0.615**	0.050	−0.342	−0.174	0.687
DIAS	**0.624**	0.063	−0.291	−0.049	−0.721
WT	0.252	**0.616**	0.373	0.642	0.078
HT	−0.117	**0.694**	0.141	−0.695	−0.051
Eigenvalue, λ	2.1894	1.5382	0.6617	0.4485	0.1621
Square root, $\sqrt{\lambda}$	1.48	1.24	0 .81	0.67	0.40
$0.6/\sqrt{\lambda}$	0.40	0 .48	0.74	0.90	1.50
Interpretations	BPtotal	SIZE	AGE	OVERWT	BPdiff

Table 4.
Loadings corresponding to correlations >0.6 are boldface.

No. of PCs, k	λ_k	$\lambda_k > 1$?	$\ln \lambda_k$	$N \ln \lambda_k$	for BIC: $N \ln \lambda_k > -4.61$?	for AIC: $N \ln \lambda_k > -2$?
1	2.19	Yes	0.78	78.36	Yes	Yes
2	1.54	Yes	0.43	43.06	Yes	Yes
3	0.66	No	−0.41	−41.29	No	No
4	0.45	No	−0.80	−80.18	No	No
5	0.16	No	−1.82	−181.95	No	No

Table 5.
Estimating the number of PCs by various methods.

this cutoff value are in boldface. (The cutoff point of 0.6 is somewhat arbitrary; one might use, for example, a cutoff of 0.5.)

One can also focus on the pattern of loadings within the different PCs for the interpretation of the PCs. To reiterate this process and the interpretations, we have the following:

PC1: SYS and DIAS have large loadings with the same sign; we interpret PC1 as BPindex, or BPtotal.

PC2: WT and HT have large loadings with the same sign; we interpret PC2 as the man's SIZE.

PC3: Only AGE has a large loading, so we interpret PC3 simply as AGE.

PC4: WT and HT have large loadings with opposite signs; we interpret PC4 as OVERWEIGHT.

PC5: SYS and DIAS have large loadings with opposite signs; we interpret PC5 as BPdrop.

We continue to marvel at how readily interpretable the PCs are. This simplicity is attained even without using a factor analysis model and using rotation to simplify the pattern of the loadings.

4.4 Employing the criteria in the example

To compare and contrast the methods, **Table 5** shows the eigenvalues and the results according to the various criteria for deciding on the adequate number of PCs. According to the rule based on the average eigenvalue, the dimension is retained if its eigenvalue is greater than 1 (when working in terms of the correlation matrix). For BIC, the kth PC is retained if

$$n \ \ln \ \lambda_k > -a_n, \tag{43}$$

where $a_n = \ln n$. Here, $n = 100$ and $\ln n = \ln 100$, approximately 4.61. For AIC, the kth PC is retained if $n \ln \lambda_k > -2$. In this example, the methods agree on retaining $k = 2$ PCs.

We feel that we should remark that, though it is the case that two PCs are suggested, the fourth and fifth PCs do have simple and interesting interpretations. It is just that they do not improve the fit very much. The third PC is essentially a single variable, age.

5. Discussion

The focus here has been on determining the number of dimensions needed to represent a complex of variables adequately. The algebraic solution devolves upon the analysis of properties of the covariance matrix of the variables, especially through its eigensystem.

5.1 Regression on principal components

Next, we consider applying principal component analysis in the context of *multiple regression*. In this context, there is, of course, a response variable Y and explanatory variables $X_1, X_2, \ldots, X_p$. One may transform the Xs to their principal components, as this may aid in the interpretation of the results of the regression. In addition, the number of significant regression coefficients may be decreased. In such *regression on principal components* (see, e.g., [10]), however, one should not necessarily eliminate the principal components with small eigenvalues, as they may still be strongly related to the response variable.

The value of the Bayesian information criterion for Model k is

$$\mathrm{BIC}_k = -2LL_k + m_k \ln n, \tag{44}$$

for alternative models indexed by $k = 1, 2, \ldots, K$, where LL_k is the maximum log likelihood for Model k, that is, $LL_k = \max \ln L_k$ and m_k is the number of independent parameters in Model k. For linear regression models with Gaussian-distributed errors, $-2LL_k = \text{Const.} + n \ln \mathrm{MSE}_k$ and so BIC takes the form

$$\mathrm{BIC}_k = n \ln \mathrm{MSE}_k + m_k \ln n, \tag{45}$$

where here MSE_k is the maximum likelihood estimate (MLE) of the mean squared error (MSE) of Model k, with divisor n, of the error variance.

The total number of subsets of p things is 2^p. Therefore, with p explanatory variables, there are 2^p alternative models—“subset regressions”—(including the model where no explanatory variables are used and the fitted value of Y is simply $\bar{y}$). For example, if there are three Xs, the eight subsets are X_1 alone, X_2 alone, X_3 alone, $(X_1, X_2), (X_1, X_3), (X_2, X_3), (X_1, X_2, X_3)$, and the empty set. It would usually seem to be expedient to evaluate all 2^p regression models—regressions on all 2^p subsets of principal components, using adjusted R-square, AIC, and/or BIC rather than reducing the number of models considered by regressing on only a few principal components. That is, in the context of regression on principal components, it is probably wise *not* to reduce the number of principal components, for, as stated above, it is conceivable that some principal components with small eigenvalues may nevertheless be important in explaining and predicting the response variable.

5.2 Some related recent literature

Other researchers have considered the problem of the choice of the number of principal components. For example, Bai et al. [11] examined the asymptotic consistency of AIC and BIC for determining the number of significant principal components in high-dimensional problems. The focus in this chapter has not necessarily been on high-dimensional problems.

Some various applications from recent literature involving choosing the number of principal components include the following. The method presented here could possibly be applied in these applications.

For example, a good book on the topic of model selection and testing, covering many aspects, is [12]. In recent years, various econometricians have examined the problems of diagnostic testing, specification testing, semiparametric estimation, and model selection. In addition, various researchers have considered whether to use model testing and model selection procedures to decide upon the models that best fit a particular dataset. This book explores both the issues with application to various regression models, including models for arbitrage pricing theory. Along the lines of model selection criteria, the book references, e.g., [8], the foundational paper for BIC.

Next, we mention some recent papers, which show applications of model selection in various research areas.

One such paper is [13], an application of principal component analysis and other methods to water quality assessment in a lake basin in China.

Another is [14], on feature selection for *classification* using principal component analysis.

As mentioned, a particularly interesting application of principal component analysis is in regression and logistic regression. We have mentioned the paper [10] on using principal component analysis in regression, taking several principal components to replace the set of explanatory variables. Another interesting application is in [15], on using principal components in *logistic* regression.

6. Conclusions

The problem of choice of the number of principal components to use to represent a complex of variables—a multivariate sample—has been considered in this chapter.

In addition to some *ad hoc* arithmetic criteria, Akaike's information criterion (AIC) and the Bayesian information criterion (BIC) have been applied here to the choice of the number of principal components to represent a dataset. The results have been compared and contrasted with *ad hoc* criteria such as retaining those principal components that explain more than an average amount of the total variance. The use of BIC is seen to correspond rather closely to the rule of retaining PCs whose eigenvalues are larger than average.

Acknowledgements

There are no further acknowledgements.

Authors' contributions

Stanley L. Sclove is the sole author.

Funding

There was no funding other than the author's usual salary at the university.

Competing interests

There are no competing interests.

Availability of data and material

The source of data used is a book that is referenced and available.

Abbreviations

AIC	Akaike's information criterion
BIC	Bayesian information criterion
DIAS	diastolic blood pressure
HT	height
LC	linear combination
LL	maximum log likekihood
MLE	maximum likelihood estimate
MSE	mean squared error
PC	principal component
SYS	systolic blood pressure
WT	weight

Author details

Stanley L. Sclove
University of Illinois at Chicago, Chicago, Illinois, USA

*Address all correspondence to: slsclove@uic.edu

References

[1] Sclove SL. Application of model-selection criteria to some problems in multivariate analysis. Psychometrika. 1987;**52**(1987):333-343. DOI: 10.1007/BF02294360

[2] Sclove SL. Principal components. In: Darity WA editor. International Encyclopedia of the Social Sciences, 2nd edition. Detroit, USA: Macmillan Reference

[3] Anderson TW. An Introduction to Multivariate Statistical Analysis. 3rd ed. New York, NY: Wiley; 2002

[4] Johnson RJ, Wichern DW. Applied Multivariate Statistical Analysis. 6th ed. Upper Saddle River, NJ: Pearson; 2008

[5] Akaike H. Information theory and an extension of the maximum likelihood principle. In: Petrov BN, Csáki F, editors. 2nd International Symposium on Information Theory, Tsahkadsor, Armenia, USSR, September 2-8, 1971. Budapest: Akadémiai Kiadó; 1973. pp. 267-281 Republished in Kotz S, Johnson NL editors. Breakthroughs in Statistics, I. Berlin, Germany: Springer-Verlag;1992. pp. 610–624

[6] Akaike H. A new look at the statistical model identification. IEEE Transactions on Automatic Control. 1974;**19**(6): 716-723. DOI: 10.1109/TAC.1974.1100705

[7] Akaike H. Prediction and entropy. In: Atkinson AC, Fienberg SE, editors. A Celebration of Statistics, Springer. NY: New York; 1985. pp. 1-24

[8] Schwarz G. Estimating the dimension of a model. The Annals of Statistics. 1978;**6**:461-464 Available from: http://www.jstor.org/stable/2958889

[9] Dixon WJ, Massey FJ Jr. Introduction to Statistical Analysis. 3rd ed. New York: McGraw-Hill; 1969

[10] Massy WF. Principal components regression in exploratory statistical research. Journal of the American Statistical Association. 1965;**60**(309): 234-256. DOI: 10.1080/01621459.1965.10480787

[11] Bai Z, Choi KP, Fujikoshi Y. Consistency of AIC and BIC in estimating the number of significant components in high-dimensional principal component analysis. The Annals of Statistics. 2018;**46**(3): 1050-1076. DOI: 10.1214/17-AOS1577

[12] Bhatti MI, Al-Shanfari H, Hossain MZ. Econometric Analysis of Model Selection and Model Testing. Oxfordshire, England, UK: Routledge; 2017

[13] Xu S, Cui Y, Yang C, Wei S, Dong W, Huang L, et al. The fuzzy comprehensive evaluation (FCE) and the principal component analysis (PCA) model simulation and its applications in water quality assessment of Nansi Lake Basin, China. Environmental Engineering Research. 2021;**26**(2):222-232

[14] Omuya EO, Okeyo GO, Kimwele MW. Feature selection for classification using principal component analysis and information gain. Expert Systems with Applications. 2021;**174**: 114765

[15] Aguilera AM, Escabias M, Valderrama MJ. Using principal components for estimating logistic regression with high-dimensional multicollinear data. Computational Statistics and Data Analysis. 2006;**50**(8): 1905-1924

Chapter 9

Spatial Principal Component Analysis of Head-Related Transfer Functions and Its Domain Dependency

Shouichi Takane

Abstract

In this chapter, the Principal Component Analysis (PCA) was adopted to spatial variation of Head-Related Transfer Function (HRTF) or its corresponding inverse Fourier Transform, called Head-Related Impulse Response (HRIR), in order to compactly represent their spatial variation. This is called the Spatial PCA (SPCA). The SPCA was carried out for a database of HRTFs in all directions by selecting the domain as one of the HRIRs, the complex HRTFs, the frequency amplitudes of HRTFs, log-amplitudes of HRTFs, and complex logarithm of HRTFs. The minimum phase approximation was incorporated for the frequency amplitudes and log-amplitudes of HRTFs. Comparison of the accuracies in both time and frequency domains taking into account their influence on subjective evaluation showed that the log-amplitudes and complex logarithm of HRTFs are suitable for the SPCA of HRTFs.

Keywords: spatial principal component analysis, head-related transfer function, head-related impulse response, domain, compact representation

1. Introduction

1.1 Head-related transfer function (HRTF)

Head-Related Transfer Function (HRTF) is defined as an acoustic transfer function from sound acquired at a center point when a listener is absent to that acquired at the listener's ear [1] in a free field (a field without any reflection). A sample of its illustration is depicted in **Figure 1**. As in **Figure 1(a)**, a microphone is located at the center of a subject's head with the subject absent. The output $Y_{\mathrm{A}}(z)$ is obtained as the response to the input $X(z)$ by using the z-transform as follows:

$$Y_{\mathrm{A}}(z) = M(z) \cdot H_{\mathrm{A}}(z) \cdot S(z) \cdot X(z), \tag{1}$$

where $M(z)$ and $S(z)$ are system functions corresponding to the microphone and loudspeaker, respectively. As in **Figure 1(b)**, the same microphone is located at the

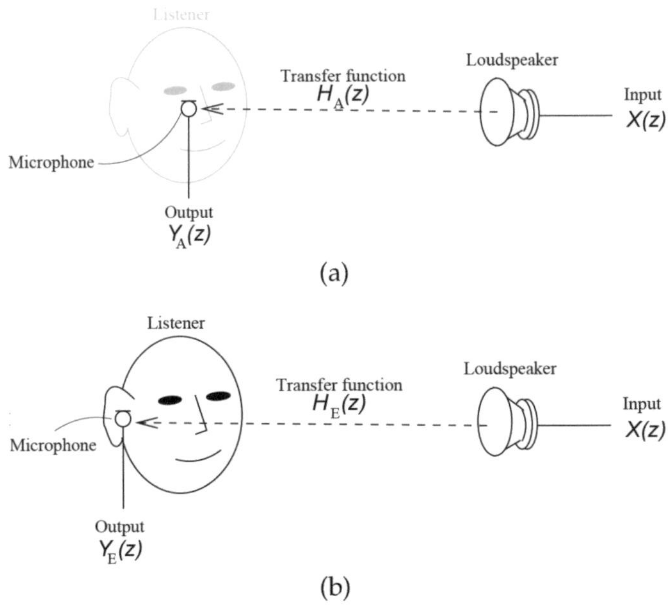

Figure 1.
Definition of head-related transfer function. (a) Obtaining the response $Y_A(z)$ at the center of a subject's head with the subject absent. (b) Obtaining the response at the ear $Y_E(z)$.

subject's ear. The output $Y_E(z)$ is also obtained as the response to the same input $X(z)$ fed to the same loudspeaker as follows:

$$Y_E(z) = M(z) \cdot H_E(z) \cdot S(z) \cdot X(z). \tag{2}$$

the z-transform of HRTF, $H(z)$, is acquired from $Y_A(z)$ and $Y_E(z)$ as follows:

$$H(z) = \frac{Y_E(z)}{Y_A(z)} = \frac{H_E(z)}{H_A(z)}. \tag{3}$$

Computation of Eq. (3) eliminates the system functions of $M(z)$ and $S(z)$ when the same microphone and loudspeaker are used for the acquisition of the HRTF, except the case that either of these system functions has zeros. The HRTF is obtained as $H(z)|_{z=\exp(j\omega)}$ where j is imaginary unit, $\omega = 2\pi f$ is the angular frequency and f is the frequency. Time domain representation (impulse response) corresponding to $H(z)$ is called as the Head-Related Impulse Response (HRIR).

The HRTF varies due to the sound source position and has strong individuality in both objective and subjective senses. Therefore a set of HRTFs is ideally acquired individually in all sound source directions. While a study considering the efficient sampling scheme of the HRTF measurement exists [2], data size of such set of HRTFs may become numerous. There also exist many datasets involving the HRTFs (HRIRs) of multiple subjects in multiple sound source directions [3–6], but the individualization using these datasets seems difficult.

1.2 Virtual auditory display (VAD) utilizing head-related transfer functions

Virtual Auditory Displays (VADs), which is a device or an equipment for presentation of an audition in certain sound field to a listener, have been developed since

1990s [7]. On the other hand, the primitive form of the VADs was proposed in 1960s [1] and Morimoto *et al.* applied the theory into practice in 1980 [8]. Some of the VADs are known to be based on the synthesis of transfer functions involving the Head-Related Transfer Functions (HRTFs). They require the real-time processing on their variation due to the movement of the listener and/or the sound sources. Takane *et al.* proposed a theory of VAD named ADVISE (Auditory Display based on VIrtual SpherE model) [9], and reported an elemental implementation of the VAD based on ADVISE [10]. The listener's own HRTFs in all directions are ideally essential in order to carry out the synthesis. Moreover, various implementations of VADs exist based on the synthesis of binaural sound signals using the HRTFs, [7, 11–13]. Taking into account a set of HRTFs acquired for an individual in all directions, its data size must be as compact as possible with their synthesis accuracy achieved to some extent.

A possible approach to the compact representation for spatial variation of an individual HRTF is modeling. Haneda *et al.* proposed the Common Acoustical-Pole and Zero (CAPZ) model [14]. In the CAPZ model, it is assumed that the poles in HRTFs are independent of sound source positions while their zeros are dependent on them. They indicated that the spatial variation of the HRTFs of a dummy head was modeled in acceptable accuracy. Based on this model, Watanabe *et al.* proposed the interpolation method and this method showed good interpolation accuracy [15]. The CAPZ model is useful for the compact representation of the HRTFs since the source-position-independent poles makes the total number of coefficients for the representation of the HRTFs with their spatial variation. The data amount decreased by using the CAPZ model, however, is up to 50% relative to the case that all HRTFs in all directions are represented by the FIR filters with fixed length.

1.3 Head-related transfer functions and principal component analysis

Another promising method for the compact representation of HRTFs is the Principal Component Analysis (PCA) [16, 17]. In some studies, the PCA has an alternative name, the spatial feature extraction method [18–20]. Both have their theoretical basis on the PCA or the Singular Value Decomposition (SVD). In these researches, the spatial variation of HRTFs is modeled by using small number of principal components or eigenvectors. Xie called the PCA adopted to the dataset(s) of HRTFs the Spatial PCA (SPCA) of HRTFs [19]. The author uses this name after Xie in this chapter. As a result of the SPCA, a HRTF in a certain direction is represented as the linear combination of relatively small number of fixed Principal Components (PCs), meaning that these components do not change according to the sound source positions against the listener. The coefficients for the PCs represent such variation. This property has a potential for effective real-time processing concerning their spatial variation due to dynamic factors. The VAD that can synthesize the HRTFs from multiple sound sources in real-time is currently available, for example by using the computational power of the Graphics Processing Units (GPUs) [21].

Many researches have been carried out on the SPCA of HRTFs [16–20], but there are some differences among these studies. One of the obvious differences is the domain to execute the SPCA. Kistler *et al.* applied the log-amplitude of the HRTF to the SPCA [17], Chen *et al.* applied the complex-valued frequency spectrum [18], and Xie applied the amplitude of the HRTF with the assumption of the minimum phase approximation [19]. On the other hand, Wu *et al.* applied the HRIRs [20]. Xie surveyed and summarized those results in his book [22]. These studies indicate that the SPCA can be successively and commonly adopted by using each domain. In contrast, the use of different domains may bring about the different properties in the results of the SPCA. If the HRTF/HRIR can be reconstructed by using the smallest number of PCs in a certain domain, the SPCA in that domain may bring about the most compact representations. There exists a study with the similar purpose. Liang *et al.* compared between the SPCA of the linear and logarithmic magnitudes of the HRTFs [23]. The conclusion of this research was that the SPCA on the linear magnitudes of the HRTFs was better than that on their logarithmic magnitudes in the reconstruction accuracy of their monaural loudness spectra. However, their used HRTFs were limited only in horizontal plane, and they only dealt with two domains with the assumption of the minimum phase approximation. Furthermore, Takane proposed the new domain for the SPCA, the complex logarithm of the HRTFs [24].

In this chapter, all domains dealt with the previous researches are picked up together and the compactness brought by the SPCA using each domain is compared.

2. SPCA of HRTFs/HRIRs

2.1 Outline

The SPCA of HRTFs/HRIRs is outlined in this section. It is a matter of course that the SPCA of HRTFs/HRIRs is based on the PCA.

1. Spatial average of a certain set of M vectors $\mathbf{g}_m$ $(m = 1, \cdots, M)$ is calculated as follows:

$$\mathbf{g}_{\mathrm{av}} = \frac{1}{M}\sum_{m=1}^{M}\mathbf{g}_m. \tag{4}$$

2. Covariance matrix, denoted as $\mathbf{R}$, is obtained by calculating the following equation:

$$\mathbf{R} = \frac{1}{M}\sum_{m=1}^{M}\left(\mathbf{g}_m - \mathbf{g}_{\mathrm{av}}\right)\cdot\left(\mathbf{g}_m - \mathbf{g}_{\mathrm{av}}\right)^{\mathrm{H}}. \tag{5}$$

It is noted that ^{H} indicates the Hermitian transpose. The size of the matrix $\mathbf{R}$ is $N \times N$, where N indicates the size of the vector $\mathbf{g}_m$.

3. The computed matrix $\mathbf{R}$ is decomposed into N pairs of PCs (eigenvectors) and eigenvalues by solving the following eigenvalue problem:

$$\mathbf{R}\cdot\mathbf{q}_k = \lambda_k\cdot\mathbf{q}_k. \tag{6}$$

As a result, a set of the eigenvalues and principal components (PCs), λ_k and $\mathbf{q}_k$ $(k = 1, \cdots, N)$, is obtained. Note that λ_k are sorted from their largest to smallest, *i. e.*, $\lambda_1 \geq \lambda_2 \geq \lambda_3 \geq \cdots \geq \lambda_N$, and the PCs are also arranged to the corresponding eigenvalues.

4. By using the matrix $\mathbf{Q}$ with $\mathbf{q}_k$ in its column vector, the weighting vector, $\mathbf{w}_m$, corresponding to the m-th vector $\mathbf{g}_m$ is calculated as follows:

$$\mathbf{w}_m = \mathbf{Q}^{\mathrm{H}}(\mathbf{g}_m - \mathbf{g}_{\mathrm{av}}). \tag{7}$$

As a result of the SPCA, the weight $\mathbf{w}_m$ is approximated by using $\mathbf{q}_1 \sim \mathbf{q}_K (1 \leq K \leq N)$ as follows:

$$(\mathbf{w}_m)_K = \mathbf{Q}_K^{\mathrm{H}}(\mathbf{g}_m - \mathbf{g}_{\mathrm{av}}), \tag{8}$$

where $\mathbf{Q}_K$ is a matrix with its column vectors $\mathbf{q}_1 \sim \mathbf{q}_K$. Length of the vector $(\mathbf{w}_m)_K$ becomes K.

In the above-mentioned procedure, the vectors and matrices are assumed to have complex values. The Hermitian transpose ^{H} is changed to the simple transpose, ^{T}, if the values of them are real. The value m reflects the sound source position, and also the individuals if the HRTFs/HRIRs of multiple individuals are used for assembling the covariance matrix.

The m-th vector, $\mathbf{g}_m$, is reconstructed by using the PCs as follows:

$$(\mathbf{g}_m)_K = \mathbf{Q}_K \cdot (\mathbf{w}_m)_K + \mathbf{g}_{\mathrm{av}}. \tag{9}$$

The computed vector, $(\mathbf{g}_m)_K$ in Eq. (9), becomes acceptable approximation when $K < N$, but this may have acceptable accuracy in principle when the Cumulative Proportion of Variance (CPV) $R^2(K)$ is close to 1.0. The CPV is defined by using the eigenvalues of the covariance matrix, λ_k $(k = 1, \cdots, N)$, as follows:

$$R^2(K) = \frac{\sum_{k=1}^{K} \lambda_k}{\sum_{k=1}^{N} \lambda_k}, \tag{10}$$

where N is the total number of components, equals to the length of the vector $\mathbf{g}_m (m = 1, \cdots, M)$.

2.2 Domains used for the SPCA of the HRTFs/HRIRs

Five domains were applied to the assembly of the covariance matrix, based on the previous researches. Kistler *et al.* applied the log-amplitudes of the HRTF [17], Xie dealt the amplitudes of the HRTF [19]. The minimum-phase approximation was assumed in these studies. Chen *et al.* dealt the complex HRTF spectrum [18], and Takane propsed the usage of the complex logarithm of HRTF [24]. At last, Wu *et al.* applied the time domain representation of the HRTFs, *i. e.*, HRIRs [20]. The domains used in these studies are treated as the modeling "domains" in this chapter. The domain "T" corresponds to the application of the HRIR to the SPCA, the domain "C" corresponds to the application of the complex HRTF, The domain "F" corresponds to that of the amplitude of the HRTF, the domain "L" corresponds to that of the logarithm of the HRTF amplitude, and the domain "CL" corresponds to that of the complex logarithm of the HRTF. This is summarized in **Table 1**.

The m-th HRIR and HRTF are respectively expressed as $\mathbf{h}_m$ and $\mathbf{H}_m$, and $\mathbf{H}_m$ is further decomposed into its amplitude and phase components as follows:

$$\mathbf{H}_m = \mathbf{A}_m \exp\{ j\Theta_m\}, \tag{11}$$

where j is the imaginary unit. The complex logarithm of $\mathbf{H}_m$ can be written as follows:

$$\log \mathbf{H}_m = \log \mathbf{A}_m + j \cdot \text{unwrap}[\Theta_m] = \mathbf{L}_m + j \cdot \text{unwrap}[\Theta_m]. \quad (12)$$

Here the imaginary part of $\log \mathbf{H}_m$, equal to the phase of the HRTF, is assumed to be unwrapped [24]. The logarithm of the HRTF amplitude vector is defined as $\mathbf{L}$, *i.e.*,$\mathbf{L}_m \equiv \log \mathbf{A}_m$. It is obvious in the domains C, F, L and CL that the frequency spectrum has the following symmetric relations:

$$H_m(k) = H_m^*(N-k), \quad \log H_m(k) = (\log H_m(N-k))^*, \quad (13)$$

$$A_m(k) = A_m(N-k), \quad L_m(k) = L_m(N-k) \quad (k = 1, \cdots, N), \quad (14)$$

where $H_m(k)$, $A_m(k)$, $L_m(k)$ are the k-th component of the vectors $\mathbf{H}_m$, $\mathbf{A}_m$ and $\mathbf{L}_m$, respectively, and * denotes the conjugate. The relations in Eqs. (13) and (14) indicate that the vector lengths can be almost halved in these domains. When the covariance matrices assembled in the domains I, C, F, L and CL are respectively denoted as $\mathbf{R}^{(\mathrm{I})}$, $\mathbf{R}^{(\mathrm{C})}$, $\mathbf{R}^{(\mathrm{F})}$, $\mathbf{R}^{(\mathrm{L})}$ and $\mathbf{R}^{(\mathrm{CL})}$, the size of $\mathbf{R}^{(\mathrm{I})}$ is $N \times N$, while those of $\mathbf{R}^{(\mathrm{C})}$, $\mathbf{R}^{(\mathrm{F})}$, $\mathbf{R}^{(\mathrm{L})}$ and $\mathbf{R}^{(\mathrm{CL})}$ are $((N/2)+1) \times ((N/2)+1)$. In this point, the domains C, F, L and CL have the advantage in the compactness compared with the domain I. On the other hand, components of $\mathbf{R}^{(\mathrm{C})}$ and $\mathbf{R}^{(\mathrm{CL})}$ are complex while those of the covariance matrices in the other domains are real.

The domains I, C, F, L and CL mean that $\mathbf{h}_m$, $\mathbf{H}_m$, $\mathbf{A}_m$, $\mathbf{L}_m$ and $\log \mathbf{H}_m$ are respectively the used domains for the SPCA. When their approximations are obtained by using the first K PCs, they are respectively denoted as $\left(\mathbf{h}_m^{(\mathrm{I})}\right)_K$, $\left(\mathbf{H}_m^{(\mathrm{C})}\right)_K$, $\left(\mathbf{A}_m^{(\mathrm{F})}\right)_K$, $\left(\mathbf{L}_m^{(\mathrm{L})}\right)_K$ and $\left(\log \mathbf{H}_m^{(\mathrm{CL})}\right)_K$. The vectors concerning the HRIR or the HRTF are calculated by using the ones estimated via the SPCA in each domain:

Domain I: From $\left(\mathbf{h}_m^{(\mathrm{I})}\right)_K$, $\left(\mathbf{H}_m^{(\mathrm{I})}\right)_K$ is obtained by using Fast Fourier Transform (FFT), and $\left(\mathbf{A}_m^{(\mathrm{I})}\right)_K$ is the amplitude corresponding to $\left(\mathbf{H}_m^{(\mathrm{I})}\right)_K$.

Domain C: From $\left(\mathbf{H}_m^{(\mathrm{C})}\right)_K$, $\left(\mathbf{A}_m^{(\mathrm{C})}\right)_K$ is obtained by computing the corresponding amplitude. After the length of the vector $\left(\mathbf{H}_m^{(\mathrm{C})}\right)_K$ is increased by applying the relation of Eq. (13), $\left(\mathbf{h}_m^{(\mathrm{C})}\right)_K$ is obtained by using inverse FFT (IFFT).

Domain F: From $\left(\mathbf{A}_m^{(\mathrm{F})}\right)_K$, $\left(\mathbf{H}_m^{(\mathrm{F})}\right)_K$ is estimated by computing the following equation with its minimum phase components $\left(\Theta_m^{(\mathrm{F})}\right)_K$ calculated using Hilbert transform [25]:

Domain	Name
HRIR	I
HRTF	C
Amplitude of HRTF	F
Log-amplitude of HRTF	L
Complex logarithm of HRTF	CL

Table 1.
Names of five domains expressing modeling conditions for the SPCA.

$$\left(\mathbf{H}_m^{(F)}\right)_K = \left(\mathbf{A}_m^{(F)}\right)_K \exp\left\{ j\left(\Theta_m^{(F)}\right)_K\right\}. \quad (15)$$

After the length of the vector $\left(\mathbf{H}_m^{(F)}\right)_K$ is increased by applying the relation of Eq. (14), the IFFT of $\left(\mathbf{H}_m^{(F)}\right)_K$ reveals the estimates of HRIR, denoted as $\left(\mathbf{h}_m^{(F)}\right)_K$.

Domain L: From $\left(\mathbf{L}_m^{(L)}\right)_K$, $\left(\mathbf{A}_m^{(L)}\right)_K$ is obtained by calculating the following equation:

$$\left(\mathbf{A}_m^{(L)}\right)_K = \exp\left\{\left(\mathbf{L}_m^{(L)}\right)_K\right\}. \quad (16)$$

Then as in the domain F, $\left(\mathbf{H}_m^{(L)}\right)_K$ is estimated by computing its minimum phase components $\left(\Theta_m^{(K)}\right)_K$ using Hilbert transform as follows:

$$\left(\mathbf{H}_m^{(L)}\right)_K = \left(\mathbf{A}_m^{(L)}\right)_K \exp\left\{ j\left(\Theta_m^{(L)}\right)_K\right\}. \quad (17)$$

After the length of the vector $\left(\mathbf{H}_m^{(L)}\right)_K$ is increased by applying the relation of Eq. (14), the IFFT of $\left(\mathbf{H}_m^{(L)}\right)_K$ reveals the estimates of HRIR, denoted as $\left(\mathbf{h}_m^{(L)}\right)_K$.

Domain CL: From $\left(\log \mathbf{H}_m^{(CL)}\right)_K$, $\left(\mathbf{H}_m^{(CL)}\right)_K$ is obtained by calculating the following equation:

$$\left(\mathbf{H}_m^{(CL)}\right)_K = \exp\left\{\left(\log \mathbf{H}_m^{(CL)}\right)_K\right\}. \quad (18)$$

The following procedure is the same as the domain C. $\left(\mathbf{A}_m^{(CL)}\right)_K$ is obtained by computing the corresponding amplitude. After the length of the vector $\left(\mathbf{H}_m^{(CL)}\right)_K$ is increased by applying the relation of Eq. (13), $\left(\mathbf{h}_m^{(CL)}\right)_K$ is obtained by using inverse FFT (IFFT).

3. Relation between number of PCs and accuracy

3.1 Conditions of analysis

A database of HRIRs of KEMAR HATS (Head And Torso Simulator) provided by Media lab. of MIT [26] was used. Liang *et al.* used the same data in their study with the similar purpose to the one in this chapter. While they used the HRTFs only in horizontal plane [23], all data in this database involving 710 pairs of HRIRs (total: 1420) with sampling frequency of 44.1 kHz were used for the investigation in this chapter. Number of HRIRs is 1420, corresponding to M in Eqs. (4) and (5).

The initial delay in each response was extracted, then 256 sample points were taken as the data for the analysis, windowing with latter half of 512-points Blackman-Harris window function adjusting its peak at that of the HRIR. The SPCA was executed by constructing the covariance matrices from the HRIRs (called as domain I), the HRTFs (domain C), the amplitude of HRTFs (domain F), the log-amplitude of HRTFs (domain L), and the complex logarithm of HRTFs (domain CL). The HRTFs/HRIRs in all directions (710 directions×2 ears) were used, and the average vector (Eq. (4)) and the covariance matrix (Eq. (5)) were calculated in each case.

3.2 Cumulative proportion of variance (CPV)

When the PCA is generally utilized for some data, the Cumulative Proportion of Variance (CPV), $R^2(K)$, defined as Eq. (10), is used for the reference indicating how much variance is covered by using the first K PCs. The change of the CPV with PC(s) in each domain was plotted in **Figure 2**. It is found out from this figure that the CPV is monotonically increased and converges to 1.0 as the number of component(s) is increased in all domain. Among five domains, the domain CL has the largest CPV value for the first PC. The domain C has the fastest increase of the CPV against the number of PC(s), and its CPV value for the domain C is almost the same as that for the domain CL when the number of components is more than 7. In contrast, the domain L has the slowest increase, especially when the number of components is more than 15. This means that relatively large number of PCs is required to cover a certain proportion of variance in data.

Reference values in the CPV, more than which all the corresponding PCs are discarded, varied in the previous studies. Kistler *et al.* set this value to 0.90 [17], Chen *et al.* and Wu *et al.* set to 0.999 [18, 20], and Xie set this to about 0.98 [19]. Direct comparison of these values is impossible since the amount of data and analyzing purposes were different in those studies, but all of these values are more than 0.9. Therefore the least numbers of components to cover four values of the CPV, 0.90, 0.95, 0.99 and 0.999 for five domains are indicated in **Table 2**. Seeing **Table 2**, the domain CL has the smallest values among five domains in all of the CPV values, and the domain C has almost the same property. This means that the variance in the spatial variation of HRTFs can be covered by using relatively small number of PCs in these domains. The domains F and L also have smaller number of PCs when the set CPV value is small. In these domains the major PCs having large corresponding eigenvalues cover the major part of variance in data. The required number of PCs increases in the domain L when the set CPV value is large. Varying the CPV values from 0.90 to 0.999, the required number of PCs becomes five to six times in the domains I, C, F and CL, while more than ten times are required in the domain L.

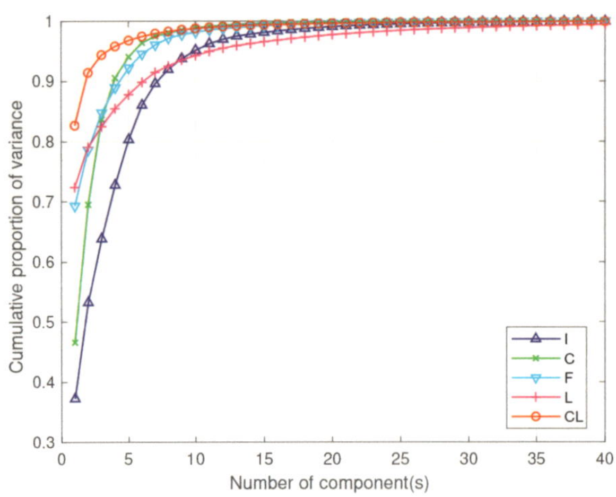

Figure 2.
Change in cumulative proportion of variance (CPV) with number of components in SPCA for five domains.

Domain	CPV			
	0.90	0.95	0.99	0.999
I	8	10	20	39
C	4	6	11	20
F	5	7	14	31
L	6	11	32	78
CL	2	4	12	39

Table 2.
The least number of PCs to cover the CPV in each case.

3.3 Reconstruction accuracy in time and frequency domains

The CPV is known to be an effective criterion for the coverage of variance with a certain number of PCs. However, comparison of the CPVs among five domains is impossible since the covariance matrices as the target for the PCA are different from each other. Therefore, the reconstruction accuracy, defined as the accuracy between the original HRTFs/HRIRs and the ones reconstructed with a certain number of PCs in five domains. In this chapter, the following two measures were computed in order to evaluate the reconstruction accuracy for the SPCA in five domains in both time and frequency domains:

Signal-to-Deviation Ratio (SDR): Signal-to-Deviation Ratio (SDR) is defined as the level difference between the energy (Euclid norm) of the original impulse response and that of the deviation:

$$\mathrm{SDR}\left[\mathbf{h}, \hat{\mathbf{h}}\right] = 10 \log_{10} \frac{||\mathbf{h}||}{\left|\left|\mathbf{h} - \hat{\mathbf{h}}\right|\right|} \quad [\mathrm{dB}], \tag{19}$$

where $\mathbf{h}$ and $\hat{\mathbf{h}}$ respectively indicate the original and the reconstructed HRIRs, $||\cdot||$ indicates the Euclid norm of the vector. The larger SDR corresponds to the closer $\hat{\mathbf{h}}$ to $\mathbf{h}$.

Spectral Distortion (SD): Spectral Distortion (SD) is defined as standard deviation in log-amplitudes of two frequency spectra, as follows:

$$\mathrm{SD}\left[\mathbf{A}, \hat{\mathbf{A}}\right] = \sqrt{\frac{1}{N_f} \sum_{k=0}^{N_f - 1} 20 \log_{10} \left| \frac{A(k)}{\hat{A}(k)} \right|} \quad [\mathrm{dB}], \tag{20}$$

where $\mathbf{A}$ and $\hat{\mathbf{A}}$ are the frequency amplitude spectrum of the original and the reconstructed responses, respectively, and $A(k)$ and $\hat{A}(k)$ are the k-th components of the vectors $\mathbf{A}$ and $\hat{\mathbf{A}}$, respectively. The value of N_f is the number of frequency bin closest to 20 kHz. The smaller SD corresponds to the closer $\hat{\mathbf{A}}$ to $\mathbf{A}$.

Calculating the SDRs for the domains F and L, the corresponding original impulse responses are ones constructed with its minimum phase approximation, which are different from the ones in the domains I, C and CL. It is noted that the SDRs in each domain were computed as how much the reconstructed impulse response differs from the desired one. Such a treatment was not related to the calculation of SDs since the SD is defined by using only magnitude of the original and the reconstructed HRTFs.

3.3.1 Changes of SDR and SD with source direction

Examples of the changes of the SDR with the source direction are plotted in **Figures 3–6. Figures 3–6** are figures for the SDR and the SD, respectively. For the elevation, elevation angles of 0°, 90° and − 90° respectively correspond to the horizontal plane, above and below the subject. The lines were plotted in these figures with 20° interval from −40° to 80°, namely the 6 lines are in each figure. For the azimuth angles, their arrangements are the same as the original data [26], *i. e.*, 0°, 90°, 180° and 270° respectively correspond to the front, right, back, and left of the subject. In each figure, number of components in each domain was set to the least value satisfying two of the CPVs in **Table 2**. The first value is 0.95 in **Figures 3** and **5**, and the second is 0.999 in **Figures 4** and **6**.

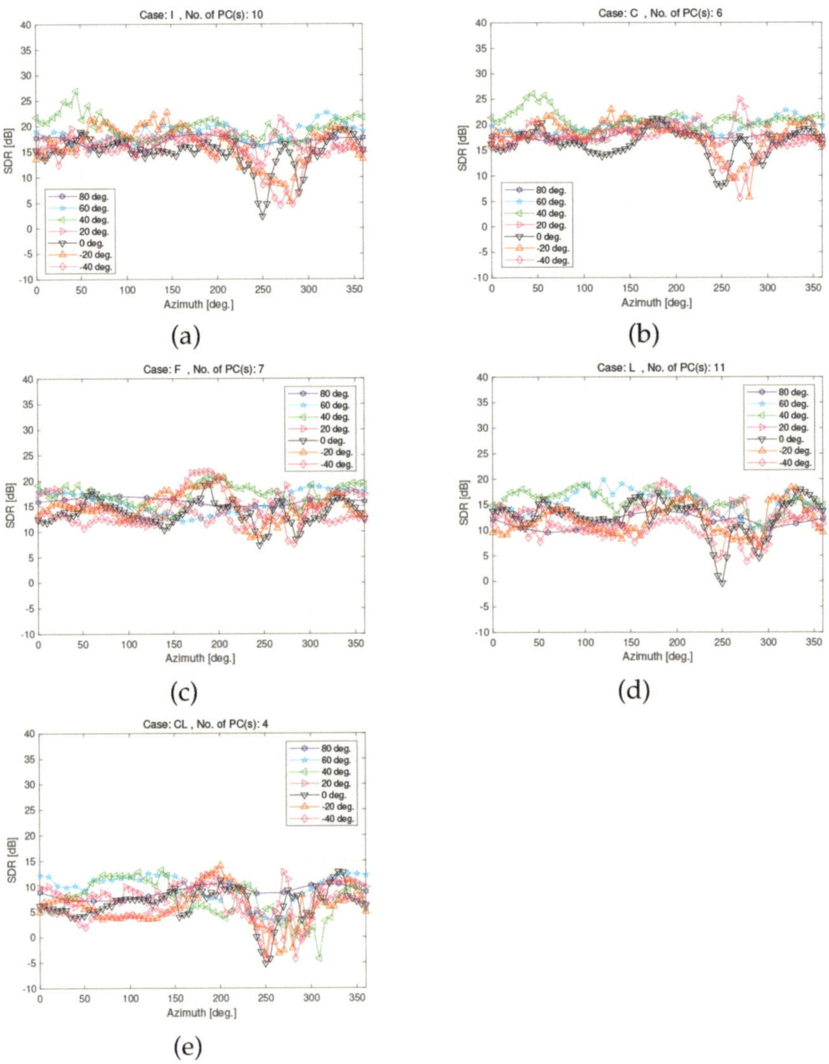

Figure 3.
Change of SDR in source azimuth at various elevation in each cases with number of PCs set to the least value achieving the cumulative proportion of variance of 0.95 in ***Table 2****. (a) Domain I (No. of PCs = 10). (b) Domain C (No. of PCs = 6). (c) Domain F (No. of PCs = 7). (d) Domain L (No. of PCs = 11). (e) Domain CL (No. of PCs = 4).*

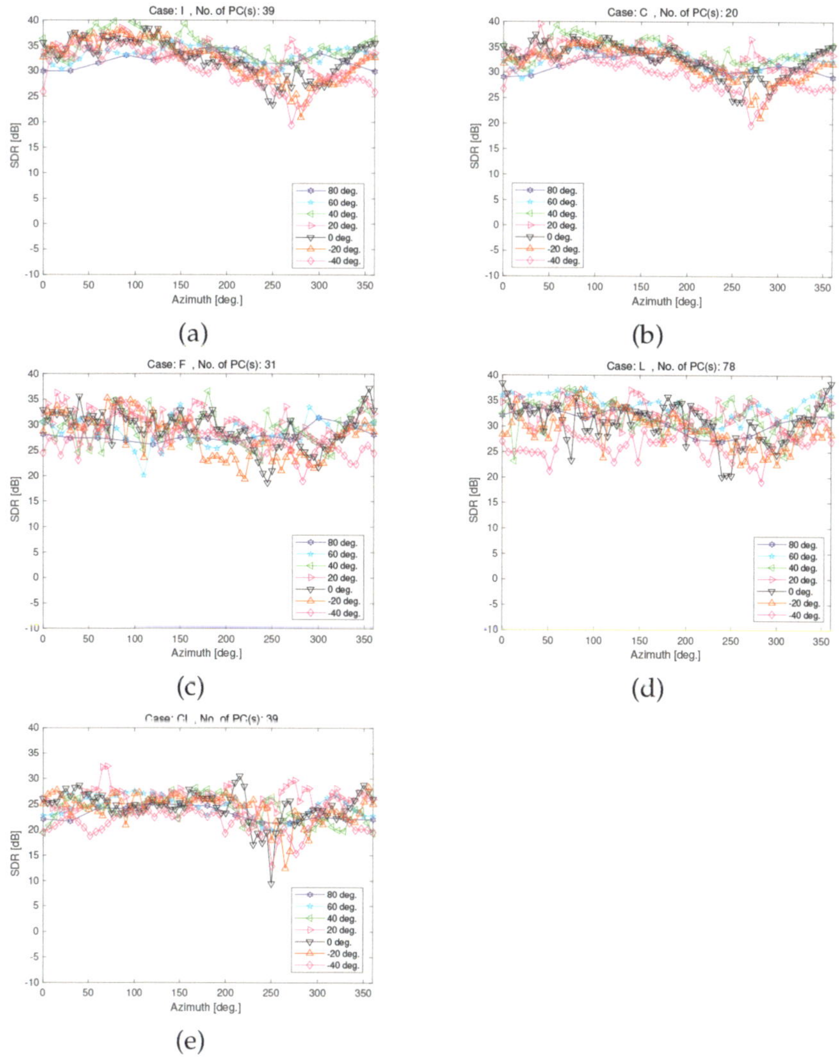

Figure 4.
Change of SDR in source azimuth at various elevation in each cases with number of PCs set to the least value achieving the cumulative proportion of variance of 0.999 in ***Table 2****. (a) Domain I (No. of PCs = 39). (b) Domain C (No. of PCs = 20). (c) Domain F (No. of PCs = 31). (d) Domain L (No. of PCs = 78). (e) Domain CL (No. of PCs = 39).*

Seeing these figures, macroscopic tendency in the change of the SDR and SD with the number of PCs is similar: the larger CPV value brings about the larger SDR and smaller SD, and these values are roughly the same when the CPV values is set equal among five domains. In contrast, it should be emphasized that the values of SD for the domains L and CL are smaller than those in the other domains, as shown in **Figure 5 (d)**, **(e)** and **6(d)**, **(e)**. The domains using the real and complex logarithm may give relatively smaller distortions in frequency domain.

Seeing the properties in time domain according to **Figures 3** and **4**, the values of SDR are gradually higher when the CPV value is larger. When the azimuth corresponds to the contralateral side (around 250° ~ 300°) and especially at the lower elevation angles (less than 0°), the relatively smaller SDR values are found out also commonly in all domains In

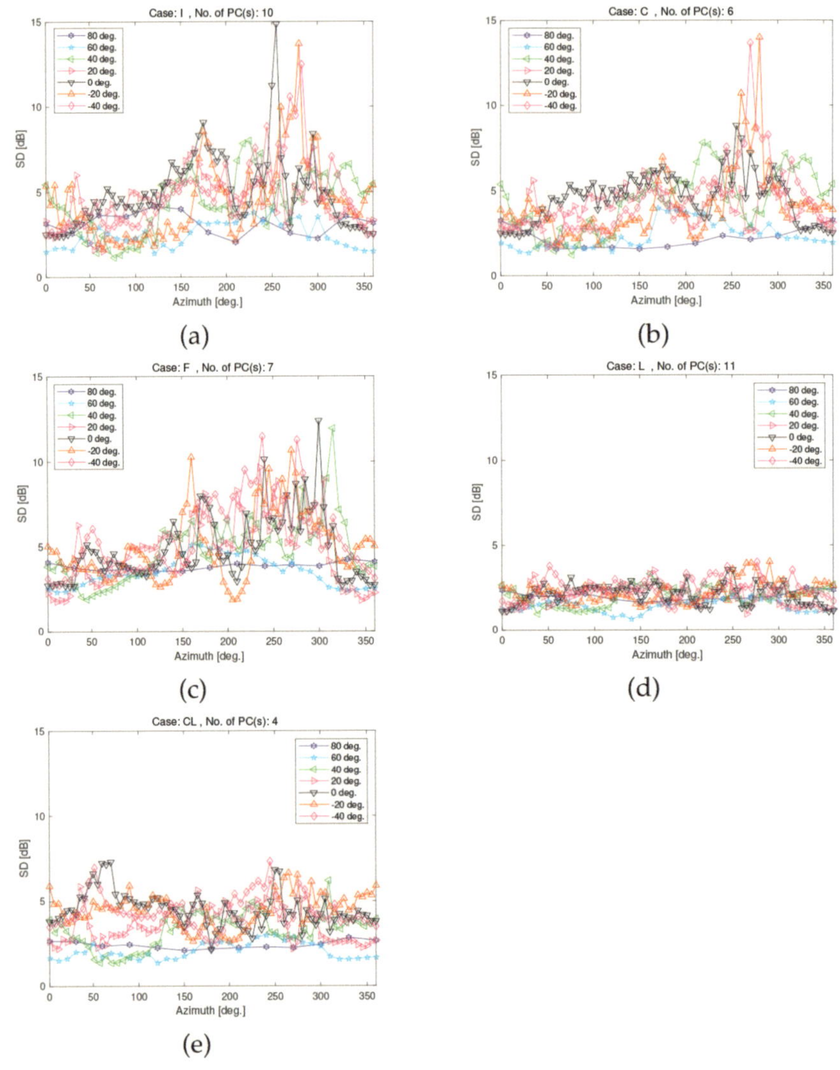

Figure 5.
Change of SD in source azimuth at various elevation in each cases with number of PCs set to the least value achieving the cumulative proportion of variance of 0.95 in ***Table 2****. (a) Domain I (No. of PCs = 10). b) Domain C (No. of PCs = 6). (c) Domain F (No. of PCs = 7). (d) Domain L (No. of PCs = 11). (e) Domain CL (No. of PCs = 4).*

these azimuths and elevation angles, the relatively larger SD values are also observed, as shown in **Figures 5** and **6**. Xie stated the same points in his articles [19, 22]. In those range of directions, the HRIRs are very small in their energy because of the subject's head making a "shadow" making the sound from the sound source hard to reach especially in the high frequency range [1]. As a result, the HRIRs in those directions are relatively difficult to be reconstructed with a small number of PCs.

3.3.2 Spatial average of SDR and SD

In order to show the macroscopic tendencies of the relation between reconstruction accuracies and domains, accuracies in time and frequency domains with number

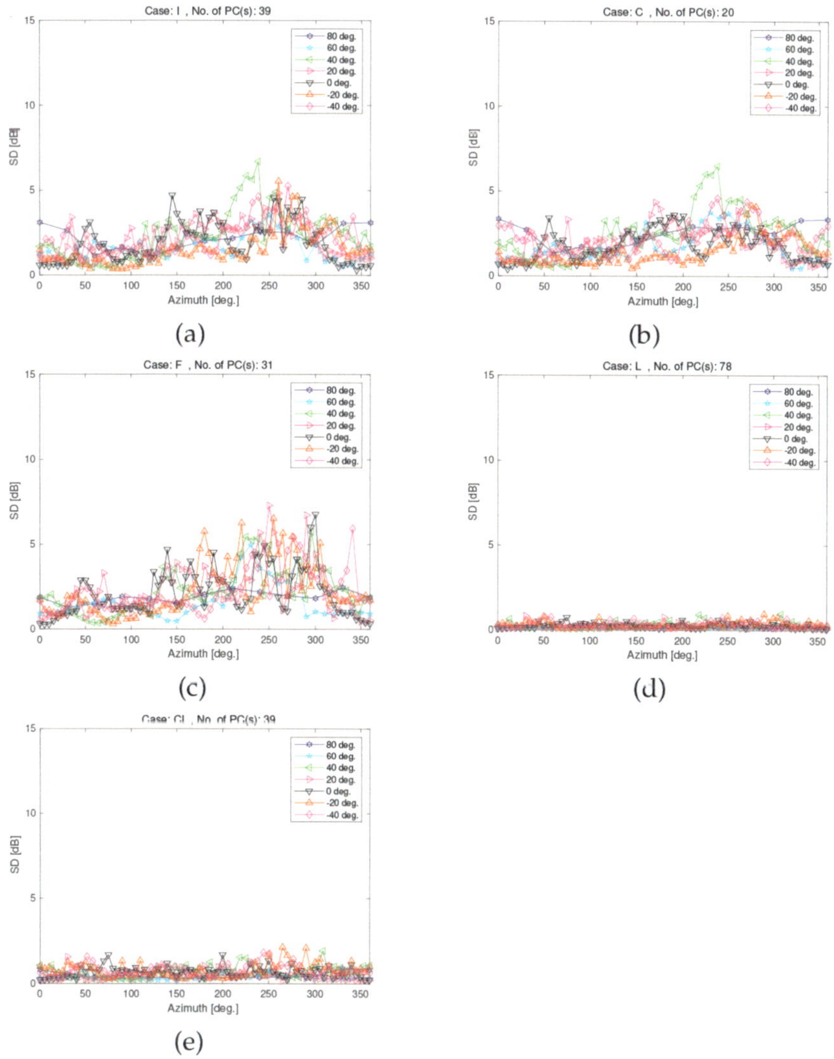

Figure 6.
Change of SD in source azimuth at various elevation in each cases with number of PCs set to the least value achieving the cumulative proportion of variance of 0.999 in ***Table 2****. (a) Domain I (No. of PCs = 39). b) Domain C (No. of PCs = 20). (c) Domain F (No. of PCs = 31). (d) Domain L (No. of PCs = 78). (e) Domain CL (No. of PCs = 39).*

of PCs set to K are computed in a certain domain X (X = I,C,F,L,CL), the overall average of SDR and SD were calculated by using the following equations:

$$\mathrm{AvSDR}(X,K) = 10\log_{10}\left\{\frac{1}{M}\sum_{m=1}^{M}10^{\mathrm{SDR}\left[\mathbf{h}_m,\left(\mathbf{h}_m^{(\mathrm{X})}\right)_K\right]/10}\right\}, \tag{21}$$

$$\mathrm{AvSD}(X,K) = \sqrt{\frac{1}{M}\sum_{m=1}^{M}\left\{\mathrm{SD}\left[\mathbf{A}_m,\left(\mathbf{A}_m^{(\mathrm{X})}\right)_K\right]\right\}^2}. \tag{22}$$

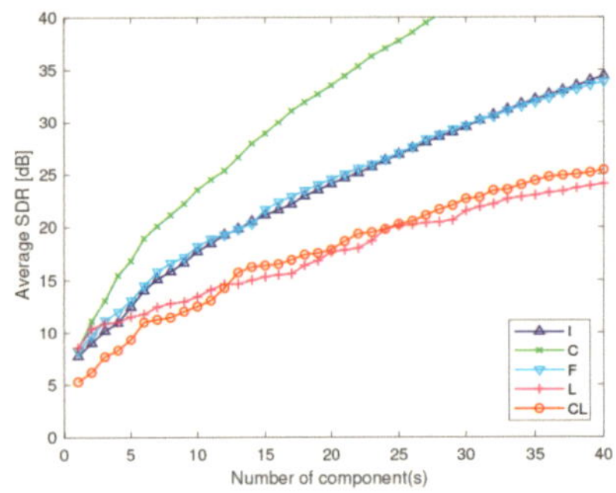

(a)

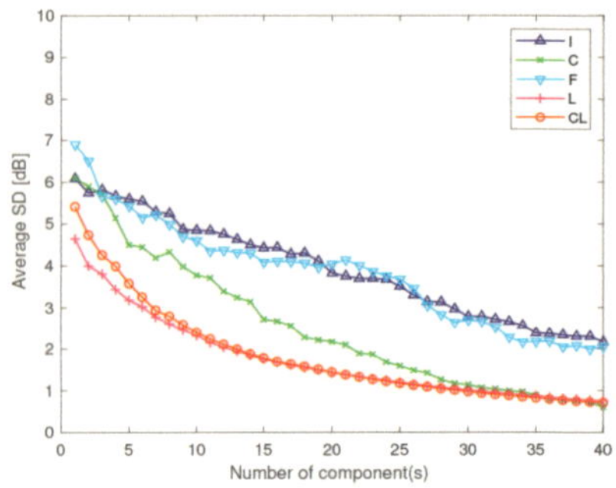

(b)

Figure 7.
Changes of (a) average SDR and (b) average SD *with number of component(s) in five domains. For SDR, five lines respectively corresponds to* $AvSDR(I,K)$, $AvSDR(C,K)$, $AvSDR(F,K)$, $AvSDR(L,K)$ *and* $AvSDR(CL,K)$, *and for* SD, *they respectively corresponds to* $AvSD(I,K)$, $AvSD(C,K)$, $AvSD(F,K)$, $AvSD(L,K)$ *and* $AvSD(CL,K)$ $(K = 1, \cdots, 40)$. *Note that average SDR is calculated with its reference to the minimum phase HRIRs in the domains* F *and* L.

Changes of the average SDR and SD with number of component(s) K in each case are plotted in **Figure 7** respectively. It is clearly found from these figures that the reconstruction accuracy improves (the larger SDR and the smaller SD) commonly in all domains as the number of PC(s) increases. This means that the CPV corresponds to the tendency of the average accuracies in both time and frequency domains. However, these values are different among five domains. Seeing **Figure 7(a)**, the largest average SDR is achieved with the domain C in most of number of PCs. The domains I and F have the similar tendency, and the domains L and CL are the lowest SDR values when the number of PC(s) is more than 5. On the other hand, as shown in **Figure 7(b)**, the domains L and CL have exceptionally the lowest SD in almost all number of PCs. The domains L and CL has the best accuracy in frequency domain. The third lowest average SD is obtained in the domain C, and the value is almost the same as in the domains L and CL when the number of PCs are relatively large (≥35).

4. Discussion

In the previous section, the SDR and SD computed in five domains were compared. It was shown that the larger number of PCs brings about the accurate reconstruction of HRTFs and HRIRs in all domains. As differences in domains in the SPCA of HRTFs and HRIRs, the average SDR has the largest value in the domain C, and the average SD has the smallest value in the domains L and CL for almost any given number of components. Since the HRTFs and HRIRs are used for the sound signals at the listener's both ears, it is essential to take their subjective evaluation into account. Most of the previous researches dealt with both the objective and subjective evaluation [17, 19, 27, 28]. However, three investigations can be found, in which the relation of the subjective evaluation to the SDR and SD values. Hanazawa *et al.* reported the results of a hearing experiment in which the relation between the accuracy of the interpolated HRIRs with their proposed method and the sound localization performance when the sound stimuli convoluted with them were presented [29]. They showed that around 6 dB in SDR had insignificant difference from the performance when the sound stimuli convoluted with the original HRIRs. Takane *et al.* also carried out an experiment to evaluated his proposed estimation method of HRIRs from the impulse responses obtained in ordinary room with reflection [30]. The results showed that the subjects did not detect significant difference between the stimuli synthesized from the estimated and the original HRIRs when the SDR between them was more than about 20 dB. Nishino *et al.* investigated the interpolation accuracy of HRTFs in the median plane, although they did not carry out the subjective evaluation. The least (best) accuracy of their interpolation method is 2 dB in average SD [31]. Considering these researches, the least numbers of PCs to achieve the average SDR more than 20 dB and the average SD less than 2 dB were checked from the results of the SDR and SD computation. The results are shown in **Table 3**. It is found out from **Table 3** that number of PCs satisfying each condition has contrastive feature. The domains I, C and F have relatively small numbers satisfying the condition of average SDR > 20 dB, meaning that these domain can reconstruct the HRIRs in relatively high accuracy with small numbers of PCs. On the contrary, The domains L and CL have relatively small numbers satisfying the condition of average SD < 2 dB, meaning that these domains can reconstruct the amplitude of HRTFs relatively high accuracy with small numbers of PCs. The domain C has the balanced property in both time and frequency domain accuracies.

It is known that the frequency-domain spectral features in the HRTFs are important for the sound localization especially in the median plane [1, 32]. Iida *et al.* proposed a parametric model of the HRTF focused on the peaks and notches in the frequency domain, and showed that the first and second lowest notches (called N1 and N2, respectively) in their frequency spectra contribute to the subjects' perceived elevation [33]. These results may state that the local frequency domain features may cause the difference in the listener's perceived direction, and the reconstruction accuracy in frequency domain must be well taken into account to avoid such potential difference. These researches support the accuracy in frequency domain is more important than that in time domain. Based on such subjective properties together with the objective properties shown in this chapter, the domain L and CL are suitable for the SPCA of the HRTFs more than the other domains. On the other hand, the domains L and CL require relatively large number of PCs to achieve the CPV values when the CPV is closer to 1. The previous researches indicate that the reconstruction with relatively small number of PCs could make the HRTFs/HRIRs without audible difference [17, 19, 27, 28], therefore it is expected that the domains L and CL can bring about the acceptable subjective evaluation with small number of PCs. The difference in the domains for the SPCA must be investigated more in detail especially based on the subjective evaluation, which is one of the future studies concerning the contents of this chapter.

Domain	Average SDR > 20 dB	Average SD < 2 dB
I	19	44
C	7	22
F	19	39
L	25	14
CL	25	14

Table 3.
The least number of components to achieve average SDR > 20 dB and average SD < 2 dB in each domain.

5. Summary

In this chapter, the SPCA of the HRTFs was introduced, and its dependency on the domains, in which the covariance matrices are calculated, was investigated. The following points are the summary of the findings in this chapter:

- The SPCA can be carried out commonly for all domains, *i. e.*, the HRIRs (domain I), the (complex) HRTFs (domain C), the amplitude spectrum of HRTFs (domain F), logarithm of the amplitude spectrum of HRTFs (domain L), and the complex logarithm of HRTFs (domain CL).

- For the domains except the domain I, the covariance matrices can be sized down to about 1/4 of the covariance matrix assembled for the domain I, according to the symmetric property of the frequency spectrum.

- The domains I, C and F have relatively small numbers of PCs in order to achieve high time domain accuracy.

- The domains L and CL have relatively small numbers of PCs in order to achieve high frequency domain accuracy.

- Considering the influence on the subjective evaluation of the reconstructed HRTFs/HRIRs with their SPCA, the domains L and CL, bringing about relatively high accuracy in frequency domain, are more suitable for the SPCA of the HRTFs.

Author details

Shouichi Takane
Tohoku Bunka Gakuen University, Sendai, Japan

*Address all correspondence to: takane@ait.tbgu.ac.jp

References

[1] Blauert J. Spatial Hearing. Cambridge, Massachusetts, USA: MIT Press. 1983

[2] Zhang W, Zhang M, Kennedy RA, Abhayapala TD. On high-resolution head-related transfer function measurements: An efficient sampling scheme. IEEE Transductions on Audio, Speech & Language Processing. 2012;**20**(2):575-584

[3] Algazi VR, Duda RO, Thompson DM. The CIPIC HRTF database. In: Proceedings of the 2001 IEEE Workshop of the Applications of Signal Processing to Audio and Acoustics; New Platz. 2001 Cat No. 01TH8575 (5 pages)

[4] Watanabe K, Iwaya Y, Suzuki Y, Takane S, Sato S. Dataset of head-related transfer functions measured with a circular loudspeaker array. Acoustical Science & Technology. 2014;**35**(3): 159-165

[5] Bomhardt R, Klein MF, Fels J. A high-resolution head-related transfer function and three-dimensional ear model database. Proceedings of Meetings on Acoustics. 2016;**29**:050002

[6] Brinkmann F, Dinakaran M, Pelzer R, Grosche P, Voss D, Weinzierl S. A cross-evaluated database of measured and simulated HRTFs including 3D head meshes, anthropometric features, and headphone impulse responses. Journal of the Audio Engineering Society. 2019; **67**(9):705-718

[7] Begault DR. 3-D Sound for Virtual Reality and Multimedia. Cambridge, Massachusetts, USA: AP Professional. 1994

[8] Morimoto M, Ando Y. On the simulation of sound localization. Journal of the Acoustical Society of Japan (E). 1980;**1**(3):167-174

[9] Takane S, Suzuki Y, Miyajima T, Sone T. A new theory for high definition virtual acoustic display named ADVISE. Acoustical Science & Technology. 2003; **24**(5):276-283

[10] Takane S, Takahashi S, Suzuki Y, Miyajima T. Elementary real-time implementation of a virtual acoustic display based on ADVISE. Acoustical Science & Technology. 2003;**24**(5): 304-310

[11] Otani M, Hirahara T. A dynamic auditory display: Its design, performance, and problems in HRTF switching. In: Proceedings of Japan-China Joint Conference on Acoustics; 4–6 June 2007; Sendai: Acoustical Society of Japan/Acoustical Society of China. 2007. SS-1-3

[12] Yairi S, Iwaya Y, Suzuki Y. Estimation of detection threshold of system latency of virtual auditory display. Applied Acoustics. 2007;**68**(8): 851-863

[13] Miller JD. Slab: A software-based real-time virtual acoustic environment rendering system. In: Proceedings of the 7th International Conference on Auditory Display; 29 July-1 August; Espoo: International Conference on Auditory Display. 2001. pp. 279-280. https://smartech.gatech.edu/handle/1853/50648

[14] Haneda Y, Makino S, Kaneda Y, Kitawaki N. Common acoustical-pole and zero modeling of head-related transfer functions. IEEE Transductions on Speech and Audio Processing. 1999;7: 188-196

[15] Watanabe K, Takane S, Suzuki Y. A novel interpolation method of HRTFs based on the common-acoustical-pole

and zero model. Acta Acustica united with Acustica. 2005;**91**:958-966

[16] Martens WL. Principal components analysis and resynthesis of spectral cues to perceived direction. In: Proceedings of the International Computer Music Conference; Champaign/Urbana: International Computer Music Conference. 1987. pp. 274-281. https://hdl.handle.net/2027/spo.bbp2372.1987.040

[17] Kistler DJ, Wightmann FL. A model of head-related transfer functions based on principal components analysis and minimum-phase reconstruction. Journal of the Acoustical Society of America. 1992;**91**(3):1637-1647

[18] Chen J, Van Veen BD, Hecox KE. A spatial feature extraction and regularization model for the head-related transfer functions. Journal of the Acoustical Society of America. 1995; **97**(1):439-452

[19] Xie B. Recovery of individual head-related transfer functions from a small set of measurements. Journal of the Acoustical Society of America. 2012; **132**(1):282-294

[20] Wu Z, Chan FHY, Lam FK, Chan JCK. A time domain binaural model based on spatial feature extraction for the head-related transfer functions. Journal of the Acoustical Society of America. 1998;**102**(4):2211-2218

[21] Watanabe K, Oikawa Y, Sato S, Takane S, Abe K. Development and performance evaluation of virtual auditory display system to synthesize sound from multiple sound sources using graphics processing unit. In: Proceedings of the 21th International Congress on Acoustics; Montreal: International Commission for Acoustics. 2013 2pEAba12 (7 pages in CD-ROM)

[22] Xie B. Head-Related Transfer Function and Virtual Auditory Display. Second ed. Plantation, Florida, USA: J. Ross Pub. 2013

[23] Liang Z, Xie B, Zhong X. Comparison of principal components analysis of linear and logarithmic magnitude of head-related transfer functions. In: Proceedings of the 2nd IEEE International Congress on Image and Signal Processing; Tianjin: IEEE. 2009. pp. 1-5. https://doi.org/10.1109/CISP.2009.530427

[24] Takane S. Spatial principal component analysis of head-related transfer functions using their complex logarithm with unwrapping of phase. In: Proceedings of the 23rd International Congress on Acoustics; 9–13 September; Aachen. 2019. pp. 3048-3055

[25] Oppenheim AV, Schafer RW. Discrete-time signal processing. third ed. Lebanon, Indiana, USA: Prentice Hall. 2010

[26] Gardner WG, Martin KD. HRTF measurements of a KEMAR. Journal of the Acoustical Society of America. 1995; **97**:3907-3908

[27] Matsui K, Ando A. Estimation of individualized head-related transfer function based on principal component analysis. Acoustical Science & Technology. 2009;**30**(5):338-347

[28] Fink KJ, Ray L. Individualization of head-related transfer functions using principal component analysis. Applied Acoustics. 2015;**87**:162-173

[29] Hanazawa K, Yanagawa H, Matsumoto M. Subjective evaluations of interpolated binaural impulse responses and their interpolation accuracies. In: Proceedings of the Spring Research Meeting of the Acoustical Society of

Japan; Tokyo: Acoustical Society of Japan. 2006. pp. 677-678 (in Japanese)

[30] Takane S, Nabatame S, Abe K, Watanabe K, Sato S. Subjective evaluation of HRIRs linearly predicted from impulse responses measured in ordinary sound field. In: Proceedings of the 39th Audio Engineering Society Japan Regional Conference; Osaka: Audio Engineering Society. 2008. pp. 1-8 (in CD-ROM)

[31] Nishino T, Kajita S, Takeda K, Itakura F. Interpolating head related transfer functions in median plane. In: Proceedings of the 1999 IEEE Workshop on Applications of Signal Processing to Audio and Acoustics; New Paltz: IEEE. 1999. pp. 17-20. https://doi.org/10.1109/ASPAA.1999.810876

[32] Asano F, Suzuki Y, Sone T. Role of spectral cues in median plane localization. Journal of the Acoustical Society of America. 1990;**88**(1):159-168

[33] Iida K, Itoh M, Itagaki A, Morimoto M. Median plane localization using a parametric model of the head-related transfer function based on spectral cues. Applied Acoustics. 2007; **68**:835-850

Chapter 10

Prediction Analysis Based on Logistic Regression Modelling

Zaloa Sanchez-Varela

Abstract

The chapter aims to show an application of logistic regression modelling for prediction analysis in the offshore industry. The different variables shown in dynamic positioning incident reports are analysed and processed using logistic regression modelling. The results of the models are then analysed, showing which data influence the loss of positioning and human errors and how the model can be interpreted. Afterwards, and based on the obtained models, operational limits can be proposed to reduce downtimes and thus improve the safety of the operations and the productivity of the offshore operations when using dynamic positioning systems.

Keywords: regression modelling, dynamic positioning, offshore, drilling, human error

1. Introduction

A dynamic positioning (DP) system is a piece of automation in which data from wind, currents and ship motions are taken from different sensors. After analysing them, a signal is sent to thrusters and rudders to compensate for those movements. This system seeks two main goals depending on the nature of the operations in progress: maintaining a given position or moving a vessel along a pre-set track.

The DP system has been in use for many decades, and its applications are primarily used in the offshore industry. The complexity and high accuracy requested for the different offshore operations make the dynamic positioning system a valuable tool for this sector.

However, rarely does such a sophisticated automated system always performs smoothly. The study of the incidents reported by vessels is vital to discover any failures that could be corrected and to improve the safety of DP operations.

The International Marine Contractors Association (IMCA) is, without any doubt, one of the most prolific authors to the cause of safety in DP operations. They have published different recommendations to the industry, along with guidelines for operations, sensors, and personnel. It is also important to mention the collection of DP incidents that IMCA has published since 1994. The high volume of DP incidents reported anonymously, and carefully published by IMCA, has been the base of this research.

In this chapter, the focus is set on the research of the incidents reported to IMCA from 2011 until 2015. During this period, the reports presented by IMCA were of the event-tree type, showing information regarding water depth, the configuration of the DP system and meteorological information. Before this period, the event trees lacked some of these data; and since 2015, the reports presented were just a sample and did not include all the incidents reported.

However, the reports collected can contribute to understanding the common patterns that can be found in the different incidents, thus finding an interpretation that could help improve the safety of these operations. Periods of downtime can mean a considerable amount of money loss. In some cases, the incident can lead to catastrophic consequences, leading to the loss of the ship and even pollution of the environment.

In this context, the research presented in this chapter aims to propose a mathematical model using logistic regression, which could help predict under which conditions an incident can occur. Furthermore, the condition of the incident can be determined beforehand, and as such, the likelihood of having an incident ending in an excursion can be modelled.

At the same time, and knowing that perhaps the human error is the easiest of all mistakes to correct, the model will be determined once again taking into account whether a human error was the cause of the incident or not, and the resulting models will be compared.

2. State of the art

2.1 DP elements

Any DP system always has seven segments or elements, namely: controller, DP console, Dynamic Positioning Operator (DPO), position reference systems (PRS), motion reference units (MRU), propulsion and power supply. Each element will be described in this section.

The central element is the controller, composed of computers or processors, which sets a two-way communication with all other DP elements via the vessel network.

The system is controlled with the help of the DP console, which contains operational controls, buttons, screens and a manual joystick.

The DP console is controlled by a Dynamic Positioning Operator (DPO) who should be fully certified to conduct DP operations.

To acquire information on the position of the rig, PRS are used. In DP drilling operations, several PRS provide additional accuracy [1]. Usually, a drilling rig will select dual differential global navigation satellite system (DGNSS) and hydro-acoustic positioning references (HPR), usually of the long-baseline type. Taut wires are only used in shallow waters, as they are not available in deep water [2].

The motions of the vessel are monitored with different sensors. The yawing is monitored with the help of one or more gyrocompasses which send information about the heading. Different MRU help send information about surge and sway.

The wind and current are also monitored for course and speed, and this information is sent to the controller. There are different wind sensors in different positions onboard the rig to avoid errors due to windscreens, turbulences provoked by structures, and other obstacles.

With all this information, the controller can predict the movement on the vessel and send a proper command to the propellers and thrusters (pitch, revolutions per minute, azimuth, rudder angle) to counter-rest the forces and maintain the rig in the desired position.

A vital part of the DP system is the power supply. Diesel, alternators, switchboards, cabling, propulsion motors and power-management form part of the power system related to the DP operations [3].

2.2 The use of dynamic positioning systems in drilling operations

Drilling operations take place over a wellhead. The primary purpose of the DP system is to maintain the position of the drilling vessel so that the riser/stack angle containing the drill string is close to zero, compensating for currents or tidal flow if necessary [4]. This angle is the one measured between the riser (on the top) and the wellhead or lower marine riser package (LMRP) [3]. This function is known as riser angle or riser follow mode. The DPO monitors the riser difference angle through sensors located around the LMRP. A watch circle system is created so the DPO can monitor the movements of the vessel. When the rig is moving, different levels of alarm are set to ensure the safety of the operations at all times [4].

The main risk in any DP operation is losing position (which is known in drilling operations as an excursion) during operations. Therefore, the DPO should react in a short time to correct or mitigate the consequences of this loss [5].

To maintain the position of the drilling riser, the system consists of a closed-loop control function that receives information from different sensors that measure wind, currents, heading and position. It sends a command to the propulsion units to counter rest the forces that, according to the information, tend to take the vessel out of position.

The desired position is input by the DPO, who supervises the operation in the Human Machine Interface (HMI), also known as the DP console. The DPO operator is a certified officer of the watch who has followed a training and certification scheme to cover this board position [6].

Finding which variables and in which way and measure they affect an incident having a human cause can help focus on the riskiest situations and improve the safety of drilling operations. From the results obtained, it would be possible to propose operational limits to improve the safety of drilling operations.

DP drilling incidents have been the object of different academic research. In 2011, Haibo Chen [7] published a paper where he introduced the safety of DP operations based on a barrier model. Previously the same research team had already published an article about the safety of such units [8].

The most interesting approach to the human factors in DP incidents has been proposed by Chae [9], while formal safety assessment was applied to them [10]. Dong [11] focused his research on the incidents that had taken place during offshore loading operations. Overgard [12] also researched the human element during DP incidents.

3. Objectives

There are two main objectives in this chapter.

The first main objective of this paper is to find the mathematical expression that determines the probability of an incident ending in an excursion during DP drilling operations.

With the developed model obtained from this research, drilling companies and other authorities can review their management manuals and propose some effective measures to reduce the probability of loss of position while conducting DP drilling operations. The excellent results obtained by the presented model avail the reliability of this technique.

The second main objective would be to determine whether the model remains the same when there is a human error or not, that is, to determine if the human error can alter the proposed model.

4. Methodology

The first step for applying a regression modelling technique is to have a database with different variables. The variables do not need to follow a normal distribution in the database, which is a condition for other prediction techniques.

In our example, data was gathered from the IMCA station-keeping incidents corresponding to 2011 to 2015. The cases that took place while drilling operations were in progress were selected, 50 in total.

The data described in the event tree was carefully read, and a database was developed, including the following variables, as shown in **Table 1**.

Some of these variables had to be treated to be used in this research. Thus, the following variables were obtained, shown in **Table 2**.

Once the database was created, some missing values were observed for some of the variables. These cases were eliminated to uniform the sample without distorting the values by performing bootstrapping.

A descriptive statistic of each variable is performed before researching the binary logistic regression models.

4.1 Binary logistic regression model

The binary logistic regression technique will provide the probability that a given variable, called the dependent variable, will have a given value based on the values of the other variables, called independent variables.

For our example, the dependent variable will be an excursion. The excursion will have a value of zero if there is no loss of position and a value of 1 if there is an excursion.

The rest of the variables will be considered independent variables. These variables can be quantitative or categorical. In our example, except for the variables water depth, percentage of thrusters online and percentage of generators online, which are all quantitative, the rest of the independent variables are categorical. Due to this, when using a statistical program, it manipulates its values internally to produce as many variables as there are categories minus one. For example, Wind sensors have five categories, and the program produces four variables: Windsensors (i), i = 1, 2, 3, 4. These new variables are dichotomic: the value 1 indicates the presence of a quality, and the value 0 its absence.

The statistical program (in our example, SPSS), considering the values for each case in the independent variables, calculates the probability of excursion for each of them. As this probability varies between 0 and 1, the closer to 0 will mean the most negligible probability of excursion, and the closer to 1 will mean a more significant probability of excursion. Thus, each case is assigned a probability p. This is important to interpret the coefficients in the regression. There has been a recodification, and no information has been lost.

Variable	Description
Year	In which year did the incident happen.
Water-depth (in metres)	Indicates the water depth at which the drilling operations took place.
Number of thrusters online	The number of thrusters that were online in the DP system.
Number of thrusters stand-by	The number of thrusters that are not online in the DP system, but which are ready to be selected at any time.
Number of generators online	The number of generators that were online in the DP system
Number of generators stand-by	The number of generators that are not online in the DP system, but which are ready to be selected at any time.
Bus tie	Whether it was open or closed.
DGNSS	The number of DGNSS systems that are selected in the DP system
HPR	The number of hydroacoustic systems that are selected in the DP system
Taut wire	The number of taut wires in use during the operations
Inertia system	The number of inertia systems in use during the drilling operations
Gyros	The number of gyros that were in use during the drilling operations
MRU	The number of MRU that were in use during the drilling operations
Wind sensors	The number of wind sensors that were in use during the drilling operations
Wind force	The force in knots of the wind blowing when the incident happened
Wind direction	The direction of the wind in degrees
Current speed	The speed of the current in knots when the incidents happened
Current direction	The direction of the current in degrees
Wave height	The height of the waves in metres
Visibility	The visibility that there was while the incident happened.
Main cause	The leading cause, as defined by the classification presented by IMCA
Secondary cause	The secondary cause, if present, as defined by the classification presented by IMCA
Excursion	Whether an excursion took place or not.

Table 1.
Variables from the database.

The choice of the variables is made by the selected method: Forward Wald. This method is based on adding or removing variables from the model by using two statistics: the score of Rao and the Wald statistic.

The score of Rao allows to compare for each independent variable Xj the null hypothesis: Ho = Bj = 0; that is, the regression coefficient B associated with the variable in the model is null. The variable that presents the minimum associated p-value provided it is always less than 0.05, for the proposed independent variable will be selected to enter the model.

Also, for the Wald statistic, the null hypothesis can be compared Ho: Bj = 0, but in this case, it is for the independent values that are already selected and have entered the model.

A variable with a p-value associated with the Wald statistic bigger than 0.1 will be eliminated, as this is by default the option of the program.

Variable	Description
Percentage of thrusters online	The number of thrusters online divided by the total number of thrusters online and stand-by.
Percentage of generators online	The number of generators online, divided by the total number of generators online and stand-by
Visibility	This variable was categorised using the following criteria: Poor when the visibility is less than two nautical miles, Moderate between 2 and 5 nautical miles, Good above five nautical miles [13].
Human cause	When either the main or secondary causes have a human origin, then 1 is inserted, for the rest of the cases, 0 is inserted indicating no human cause.
Period	The first period is from 2011 till 2013. The second period is from 2014 to 2015.

Table 2.
Variables created from existing variables.

According to the criteria exposed above, there will be several steps in which independent variables will be entered and eliminated.

At step 0, only the constant is introduced to the model. For this constant, it is essential to measure B (the regression coefficient), the estimated standard error in the estimation (SE), the Wald statistic and its degrees of freedom (df) and the associated p-value. When this p-value is less than 0.1, the constant is considered to be significant.

All the independent variables are out of the model at this step. One variable has to be selected to enter the model in step 1. The variable with the smaller p-value associated with the score of Rao, provided it is less than 0.05, will be selected. It should be considered that the variables created from a categorical variable should be considered as a whole.

If two or more variables have the same p-value, the score should then be considered, choosing the variable with the bigger score to enter the model in Step 1.

Once the variable enters the model, we should study the Wald statistic, given by:

$$Wald = (B/SE)^2 \tag{1}$$

If its p-value is above 0.1 (output value, POUT), then the corresponding variable would be eliminated (as a whole in the case of the categorical variables). It is always eliminated before the new variable is selected.

After this, another variable would be selected (or not) to enter the model in the next step. Suppose no variable can be selected due to the p-values of the score of Rao. In that case, the process is terminated, and the model is presented with a mathematical formula, given as:

$$Z = B_1X_1 + \ldots + B_qX_q + B_0 \tag{2}$$

being q the number of independent variables, and B the regression coefficients of the independent variables included in the model.

This model would explain the probability of the dependent value to be 1, that is, the possibility of an incident having a loss of position. The parameters that must be estimated are the regression coefficients B_0, B_1, ..., Bq.

The column SE presents the standard error for estimating these coefficients, which is necessary for calculating the Wald statistic.

From here, the probability p of a case having an excursion is given by:

$$p = 1/\left(1 + e^{(-Z)}\right) \tag{3}$$

So the probability p for each case can be obtained. When the value p is less than 0.5, it will indicate that the model classifies this case in the first group (not having excursion), and when the value is bigger than 0.5, then the model predicts the case to have an excursion:

Moreover, the probability of not having any loss of position is:

$$q = 1 - p \tag{4}$$

Furthermore, the relative ratio is defined as:

$$\begin{aligned} p/q &= \left(1/\left(1 + e^{(-Z)}\right)\right)/\left(1 - 1/\left(1 + e^{(-Z)}\right)\right) \\ &= \left(1/\left(1 + e^{(-Z)}\right)\right)/\left(\left(1 + e^{(-Z)} - 1\right)/\left(1 + e^{(-Z)}\right)\right) = 1/e^{(-Z)} = e^{(Z)} \end{aligned} \tag{5}$$

Then, the mean relative ratio can be obtained. According to the definition of relative ratio, the i-th incident will be more likely to occur if $P/Q > 1.0$, while another incident will be more prone to be associated with not having an excursion when this ratio $P/Q < 1.0$.

4.2 Goodness of fit

It is not enough to give the model, as the goodness of fit must be checked to decide whether the model is good or not.

We have estimated the possibility of an incident having or not an excursion, but this does not necessarily need to be real. According to the model, the case can have a more significant possibility of belonging to the first group (no excursion) and yet belong to the second group (excursion). It is a bigger problem when the probabilities are close to 0.5. In this case, there is an error, the difference between the observed probability and the estimated probability E_i = p observed$_i$ – p estimated$_i$, where p_i = can take the values 1 or 0, depending on whether the case belongs or not to the second group.

Evidencing the goodness of fit is checking how probable the obtained results for the estimated model are. It is based on comparing the number of cases that belong to the second group (excursion - yes) with the expected number if the model is valid. This expected number is the product of the total of cases in the sample by the estimated probability of belonging to the second group.

The statistic -2Log Likelihood (abbreviated -2LL) is used for this fit. When the -2LL results in low values, the likelihood is significant; the closer to zero, the bigger the likelihood.

Also, the following statistic can be used to compare the observed probabilities with the estimated from the model:

$$Z^2 = \frac{\sum_{i=1}^{n} E_i^2}{pestimated_i\left(1 - pestimated_i\right)} \tag{6}$$

They both follow a chi-square distribution with n-2 degrees of freedom under the hypothesis that the model adjusts to the observed data. It shows the percentage of correctly classified cases after the model has been defined.

When the percentage of correctly classified cases is high, it is expected to provide good results when predicting whether any incident will have an excursion or not.

5. Results

5.1 Descriptive statistics

All 42 cases were included in the analysis. There were no missing cases. 13 had a position loss from these cases, meaning 31% of the total.

5.1.1 Water depth

The mean water depth is 1409 ± 112 metres, the minimum 37 metres and the maximum 2838 metres. The distribution of this variable does not follow the normality.

5.1.2 Percentage of thrusters online

The mean usage of thrusters is 93 ± 2 per cent, with a minimum of 50% and a maximum of 100%. The distribution is not normal.

5.1.3 Percentage of generators online

The mean percentage of generators is 65 ± 3%, with a minimum of 33.33% and a maximum of 100%. This distribution is not normal.

5.1.4 DGNSS

In 3 cases (7%), there is only 1 DNGSS online, in 27 cases (64%), there are 2 DGNSS in function), in 7 cases (17%), there are 3 DGNSS online, and in 5 cases, there are 4 DGNSS working online (12%). The mean value is 2.33 ± 0.79. This distribution is not normal.

5.1.5 HPR

In 4 cases (10%), there was no HPR functioning; in 17 cases, there was 1 HPR working (40%), and in 21 cases, there were 2 HPRs in function (50%). Thus, the mean is set at 1.40 ± 0.67. This distribution is not normal.

5.1.6 Taut wire

There are 38 cases in which the taut wire is not used (91%), while in 3 cases, there is one taut wire in use (7%), and in 1 case, two taut wires were being used (2%). The mean is 0.12 ± 0.395.

5.1.7 Inertia system

In 40 cases (95%), this system is not used, while in 2 cases (5%), they are using it. The mean is then 0.05 ± 0.216. This distribution is not normal.

5.1.8 Gyros

In 1 case, only two gyros were used (2.5%), in 40 cases, three gyros were used (95%), and in 1 case, four gyros were used (2.5%). The mean value is 3 ± 0.221.

5.1.9 MRU

In 4 cases, there were 2 MRUs in use (9.5%), while in 38 cases, there were 3 MRUs in use (90.5%). The mean value is 2.90 ± 0.297.

5.1.10 Wind sensors

In 1 case, there was only one wind sensor online (2%), in 10 cases, there were two wind sensors (24%), in 26 cases, there were three wind sensors (62%), and in 5 cases, there were four wind sensors (12%). The mean is 2.83 ± 0.660.

5.1.11 Wind force

The mean wind force is 16 ± 1.88 knots, the minimum one and the maximum 55 knots metres. The distribution of this variable does not follow the normality.

5.1.12 Current speed

The mean current speed is 1.9 ± 0.23 knots. The minimum is 0.3, and the maximum is 6 knots. This distribution is not normal.

5.1.13 Wave height

The mean wave height is 1.88 ± 0.3 metres. The minimum is 0.1, and the maximum is 9.5 metres. This distribution is not normal.

5.1.14 Visibility

In 2 cases, the visibility was poor (5%), in 4 cases, it was moderate (9%), and in 36 cases, the visibility was good (86%).

5.1.14.1 Human cause

The human nature of the cause is considered when either the main cause or the secondary cause is human. There are 33 cases (78.6%) without human cause, while 9 cases (21.4%) were having a human error origin. The incidents end in an excursion in 10 cases (76.9%) when there is no human cause, and in 3 cases (23.1%) when there is a human cause. When there is not a loss of position, incidents without a human cause occur in 23 cases (79.3%), and incidents with a human cause happen in 6 cases (20.7%).

5.2 Binary logistic regression model: Dependent variable: excursion

As a previous stage, the variables are introduced in the model one by one to check their significance for explaining the answer. These variables are listed in **Table 3**.

Considering these results, the following variables are selected to enter the model: Water-depth, percentage of generators online, Wind force and Wave height.

In step 0, when the variables are not yet in the equation, the more significant (having the smallest p-value) is the percentage of generators, so this is the variable that enters the equation in step 1. After this, in step 2, the variable water-depth is also included in the equation. The different statistics can be observed in **Table 4**.

The following expression defines the model:

Causal factor	B	Wald	p-value	Odds Ratio (Exp(B))	IC 95%	
					lower	upper
Water depth	−0.001	4.645	0.031	0.999	0.998	1.000
Percentage of thrusters			Not in the equation			
Percentage of generators	0.047	8.057	0.005	1.048	1.015	1.082
DGNSS			Not in the equation			
HPR	—	0.000	1.000	—	—	—
Taut wire	—	0.000	1.000	—	—	—
Inertia system			Not in the equation			
Gyros			Not in the equation			
MRUs	−22.373	0.000	0.999	0.000	0.000	—
Wind sensors	—	5.389	0.145	—	—	—
Wind force	0.078	5.084	0.024	1.081	1.010	1.156
Force Beaufort			Not in the equation			
Wind direction			Not in the equation			
Current speed			Not in the equation			
Current direction			Not in the equation			
Wave height	0.437	3	0.081	1.549	0.947	2.532
Visibility ordinal			Not in the equation			

Table 3.
Individual results in step 1 for each independent variable when the forward (Wald) binary regression model is performed, being excursion the dependent variable.

Variables in the equation	B	S.E.	Wald	df	Sig.	Exp(B)	95% C.I.for EXP(B)	
							Lower	Upper
Water depth (m)	−0.001	0.001	4.498	1	0.034	0.999	0.998	1.000
Percentage of generators online	0.051	0.018	7.732	1	0.005	1.052	1.015	1.091
Constant	−2.641	1.378	3.676	1	0.055	0.071		

Table 4.
Variables in the equation in step 2. All variables are in step 2.

$$Z = -2.641 - 0.001 \cdot Waterdepth + 0.051 \cdot Perc.ofgenerators \tag{7}$$

The mean ratio can then be expressed as:

$$\frac{p}{q} = e^{-2.64} \cdot e^{-0.001 \cdot Waterdepth} \cdot e^{0.051 \cdot Percentageofgenerators} \tag{8}$$

Alternatively, using the Odds Ratio (column Exp(B):

$$\frac{p}{q} = e^{-2.64} \cdot 0.999^{Waterdepth} \cdot 1.052^{Percentageofgenerators} \tag{9}$$

In **Table 5**, the values obtained from the binary regression model, Z, P and P/Q, for each incident can be found.

The goodness of fit is given by the -2LL statistic and the percentage of correctly classified cases. This statistic has a value of 42.732 for Step 1 and 37.510 for Step 2. This indicates that the goodness of fit is improved in Step 2 of the model.

Out of the 42 valid cases, 29 were not ending in an excursion, while 13 had a loss of position.

In Step 1, after the variable percentage of generators was included in the equation, it was obtained that from the 29 cases without excursion, according to the model, there are 25 cases correctly classified (86.2% of the total) and that from the ones

Incident	Water-depth	Percentage of generators	*Z*	*P*	*P/Q*	*P estimated*	*P observed*	*E*
1	338	66.67	0.42117	0.603763187	1.523743	1	0	1
2	1700	50.00	−1.79100	0.142950164	0.166793	0	0	0
3	1860	50.00	−1.95100	0.124444359	0.142132	0	0	0
4	744	33.33	−1.68517	0.156412089	0.185413	0	0	0
5	750	66.67	0.00917	0.502292484	1.009212	1	0	1
6	1656	50.00	−1.74700	0.148425985	0.174296	0	0	0
7	1656	50.00	−1.74700	0.148425985	0.174296	0	0	0
8	1656	62.50	−1.10950	0.247964116	0.329724	0	0	0
9	1900	50.00	−1.99100	0.120151107	0.136559	0	0	0
10	1782	50.00	−1.87300	0.133194979	0.153662	0	0	0
11	1782	50.00	−1.87300	0.133194979	0.153662	0	0	0
12	1782	50.00	−1.87300	0.133194979	0.153662	0	0	0
13	1782	50.00	−1.87300	0.133194979	0.153662	0	0	0
14	1782	57.14	−1.50886	0.181107803	0.221162	0	0	0
15	2465	50.00	−2.55600	0.072024433	0.077615	0	0	0
16	1718	50.00	−1.80900	0.140759028	0.163818	0	0	0
17	1233	42.86	−1.68814	0.156020605	0.184863	0	0	0
18	1340	50.00	−1.43100	0.19294292	0.239070	0	0	0
19	1700	100.00	0.75900	0.681136584	2.136139	1	0	1
20	1700	100.00	0.75900	0.681136584	2.136139	1	0	1

Incident	Water-depth	Percentage of generators	Z	P	P/Q	P estimated	P observed	E
21	1250	50.00	−1.34100	0.207345657	0.261584	0	0	0
22	880	50.00	−0.97100	0.274681226	0.378704	0	0	0
23	2460	100.00	−0.00100	0.499750000	0.999000	0	0	0
24	2090	50.00	−2.18100	0.101469718	0.112929	0	0	0
25	1710	100.00	0.74900	0.678960765	2.114884	1	0	1
26	1300	50.00	−1.39100	0.199248161	0.248826	0	0	0
27	1437	50.00	−1.52800	0.178286498	0.216969	0	0	0
28	2838	37.50	−3.56650	0.027478187	0.028255	0	0	0
29	450	50.00	−0.54100	0.367954987	0.582166	0	0	0
30	315	100.00	2.14400	0.895106767	8.533503	1	1	0
31	798	50.00	−0.88900	0.291316235	0.411067	0	1	−1
32	2506	66.67	−1.74683	0.148447474	0.174326	0	1	−1
33	2118	42.86	−2.57314	0.070887218	0.076296	0	1	−1
34	38	66.67	0.72117	0.672864607	2.056838	1	1	0
35	108	66.67	0.65117	0.657274071	1.917783	1	1	0
36	959	100.00	1.50000	0.817574476	4.481689	1	1	0
37	854	100.00	1.60500	0.832716044	4.977860	1	1	0
38	1850	100.00	0.60900	0.647712655	1.838592	1	1	0
39	2100	100.00	0.35900	0.588798341	1.431897	1	1	0
40	37	100.00	2.42200	0.918489603	11.26837	1	1	0
41	1710	100.00	0.74900	0.678960765	2.114884	1	1	0
42	54	50.00	−0.145	0.463813380	0.865022	0	1	−1

Table 5.
Values obtained from the binary regression model, Z, P and P/Q, for each incident.

having a loss of position, seven are correctly classified (53.8% of the total). There are 25 + 7 = 32 cases out of 42 that are correctly classified, representing 76.2% of the studied incidents.

In Step 2, when the variable water-depth is included in the equation, for the incidents not having excursion, the number of correctly-classified cases is maintained, and for the cases with excursion, there is an improvement in the number of correctly classified cases, which are now 8 (61.5% of the total). In this second step, there are now 33 cases correctly classified, which means 78.6% of the studied incidents. There has been an evident improvement in the model with the addition of the variable water depth.

Figure 1 graphically shows the model predictions for loss of position for different values of water-depth and percentage of generators.

The relative ratio can show the prediction for loss of position for the different main causes, as shown in **Figure 2**. The dashed line allows us to appreciate better those mean values above 1, which show a higher likelihood of having a loss of position. The main causes that are more prone to end in an excursion are environmental, computer and human.

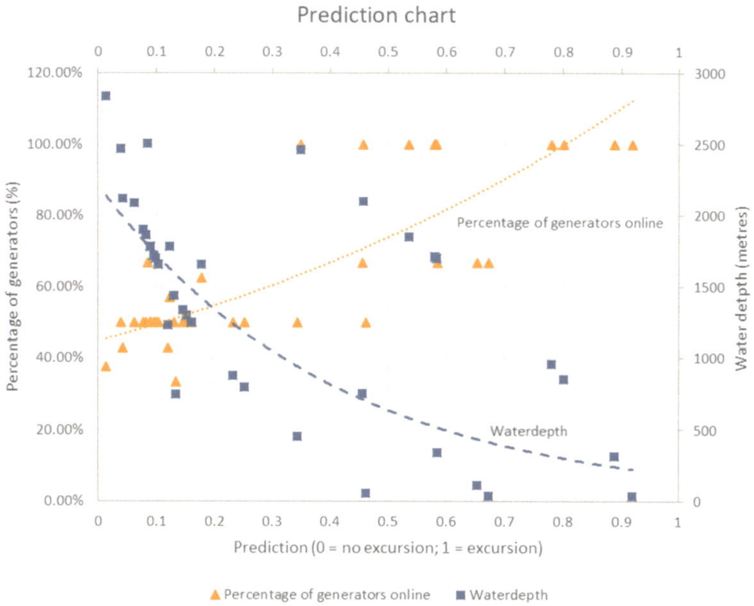

Figure 1.
Prediction chart showing the trends for wind force and percentage of generators according to the prediction model, for cases with no human cause.

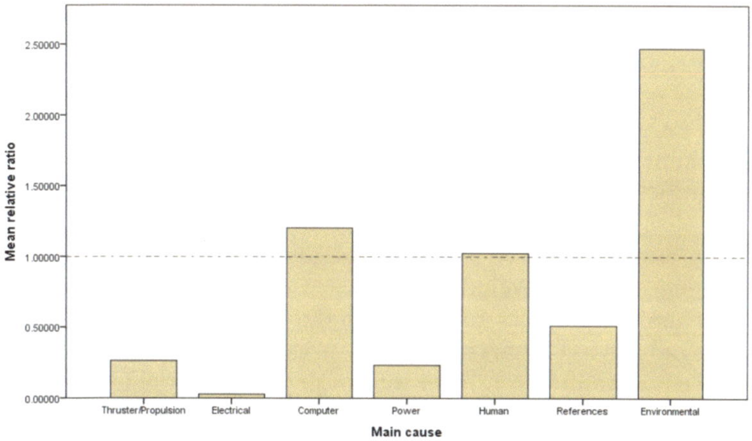

Figure 2.
Mean relative ratio for each main cause group.

The distribution of the mean relative ratio among the human cause or not of the incident is shown in **Figure 3**. The dashed line shows the value 1; above this value, the incidents are more prone to have a loss of position according to the prediction model. In this case, the incidents without a human cause have a bigger likelihood to end in a loss of position.

5.3 Model stratified by human cause

Of the 42 selected cases, 9 have a main or secondary cause with a human origin, and 33 have no evidence of human causality.

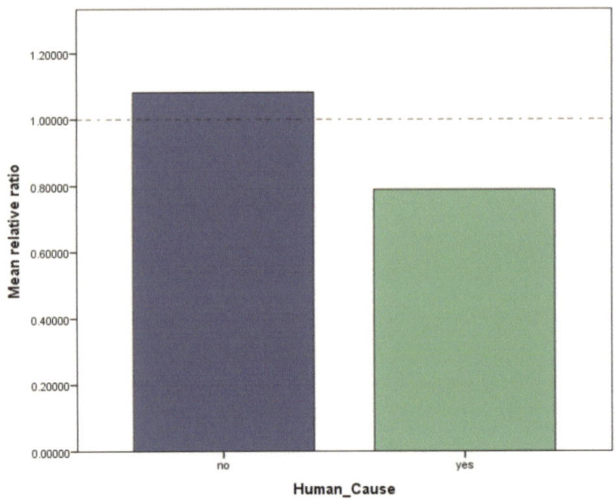

Figure 3.
Distribution of the mean relative ratio among the existence or not of a human cause for the incident.

There are no significant changes in the means of the variables when they are split into the subgroups human cause no and human cause yes, except for the variable percentage of thrusters, where it can be observed that the mean is 97.46 ± 1.48% when there is no human cause, and 74.54 ± 6.89% when there is a human cause.

5.3.1 No human cause

The 33 cases where there is no human cause are selected.

In the preliminary stage, the variables are introduced in the model one by one to check their significance for explaining the answer. The variables are presented in **Table 6**.

Considering these results, the following variables are selected to enter the model: Percentage of generators and Wind force.

In step 0, when the variables are not yet in the equation, the more significative (with less p-value) is wind force (score 7.085, p-value 0.008), so this is the variable that enters the equation in step 1. After this, in step 2, the variable water depth is also included in the equation (score 5.436, p-value 0.02). The different statistics obtained in Step 2 can be observed in **Table 7**.

The following expression defines the model:

$$Z = -6.223 + 0.051 \bullet Perc.ofgenerators + 0.12 \bullet Windforce \qquad (10)$$

The relative mean ratio of excursion can then be expressed as:

$$\frac{p}{q} = e^{-6.223} \bullet e^{0.051 \bullet Percentageofgenerators} \bullet e^{0.12 \bullet Windforce} \qquad (11)$$

Alternatively, using the Odds Ratio (column Exp(B):

$$\frac{p}{q} = e^{-6.223} \bullet 1.052^{Percentageofgenerators} \bullet 1.128^{Windforce} \qquad (12)$$

Causal factor	n	B	Wald	p-value	Odds Ratio (Exp(B))	IC 95%	
						lower	upper
Water-depth	33				Not in the equation		
Percentage of thrusters	33				Not in the equation		
Percentage of generators	33	0.039	4.757	0.029	1.04	1.004	1.078
DGNSS	33				Not in the equation		
HPR	33				Not in the equation		
Taut wire	33				Not in the equation		
Inertia system	33				Not in the equation		
Gyros	33				Not in the equation		
MRUs	33	22.547	0.000	0.999	6192653647	0.000	—
Wind sensors	33				Not in the equation		
Windforce	33	0.090	4.368	0.037	1.095	1.006	1.191
Force Beaufort	33				Not in the equation		
Wind direction	33				Not in the equation		
Current speed	33				Not in the equation		
Current direction	33				Not in the equation		
Wave height	33	0.393	2.641	0.104	1.481	0.922	2.378
Visibility ordinal	33				Not in the equation		

Table 6.
Individual results in step 1 for each independent variable when the forward (Wald) binary regression model is performed, being excursion the dependent variable and selection variable human cause = 0 (no human cause).

Variables in the equation	B	S.E.	Wald	df	Sig.	Exp(B)	95% C.I.for EXP(B)	
							Lower	Upper
Percentage of generators online	0.051	0.024	4.344	1	0.037	1.052	1.003	1.104
Windforce	0.120	0.061	3.876	1	0.049	1.128	1.001	1.271
Constant	−6.223	2.371	6.889	1	0.009	0.002		

Table 7.
Variables in the equation in step 2.

In **Table 8**, the values obtained from the binary regression model, Z, P and P/Q, for each incident can be found.

The goodness of fit is given by the -2LL statistic and the percentage of correctly classified cases. This statistic has a value of 33.453 for Step 1 and 28.147 for Step 2. This indicates that the goodness of fit is improved in Step 2 of the model.

Out of the 33 valid cases, 23 were not ending in an excursion, while ten had a loss of position.

In Step 1, after the variable wind force was included in the equation, it was obtained that from the 23 cases without excursion, according to the model, all of them were correctly classified (100% of the total) and that from the ones having a loss of

Incident	Percentage of generators	Wind force	Z	P	P/Q	P estimated	P observed	E
1	66.67	25.00	0.17717	0.544177004	1.193834	1	0	1
4	33.33	5.00	−3.92317	0.019394704	0.019778	0	0	0
5	66.67	15.00	−1.02283	0.264476518	0.359576	0	0	0
6	50.00	12.00	−2.233	0.096825971	0.107206	0	0	0
7	50.00	22.00	−1.033	0.262502905	0.355938	0	0	0
8	62.50	19.00	−0.7555	0.319624061	0.469776	0	0	0
9	50.00	10.00	−2.473	0.07777279	0.084331	0	0	0
10	50.00	4.00	−3.193	0.039429997	0.041049	0	0	0
11	50.00	19.00	−1.393	0.198929256	0.248329	0	0	0
12	50.00	15.00	−1.873	0.133194979	0.153662	0	0	0
13	50.00	12.00	−2.233	0.096825971	0.107206	0	0	0
14	57.14	18.00	−1.14886	0.24069737	0.316998	0	0	0
16	50.00	9.00	−2.593	0.069590289	0.074795	0	0	0
17	42.86	12.00	−2.59714	0.069322712	0.074486	0	0	0
18	50.00	2.00	−3.433	0.0312799	0.03229	0	0	0
19	100.00	4.00	−0.643	0.3445687	0.525713	0	0	0
20	100.00	13.00	0.437	0.607543959	1.548056	1	0	1
21	50.00	8.00	−2.713	0.0622106	0.066337	0	0	0
23	100.00	5.00	−0.523	0.372151001	0.592740	0	0	0
24	50.00	8.00	−2.713	0.0622106	0.066337	0	0	0
25	100.00	11.00	0.197	0.549091337	1.217744	1	0	1
27	50.00	6.00	−2.953	0.049594915	0.052183	0	0	0
29	50.00	16.00	−1.753	0.14766921	0.173253	0	0	0
30	100.00	50.00	4.877	0.992437784	131.2364	1	1	0
31	50.00	45.00	1.727	0.849028285	5.623757	1	1	0
32	66.67	11.20	−1.47883	0.185604206	0.227904	0	1	−1
33	42.86	18.00	−1.87714	0.132717725	0.153027	0	1	−1
34	66.67	26.00	0.29717	0.573750554	1.346044	1	1	0
35	66.67	18.00	−0.66283	0.340104178	0.515391	0	1	−1
37	100.00	55.00	5.477	0.995835559	239.1282	1	1	0
38	100.00	13.00	0.437	0.607543959	1.548056	1	1	0
39	100.00	7.00	−0.283	0.42971844	0.75352	0	1	−1
41	100.00	1.00	−1.003	0.268351995	0.366777	0	1	−1

Table 8.
Values obtained from the binary regression model, Z, P, and P/Q, for each incident without human cause.

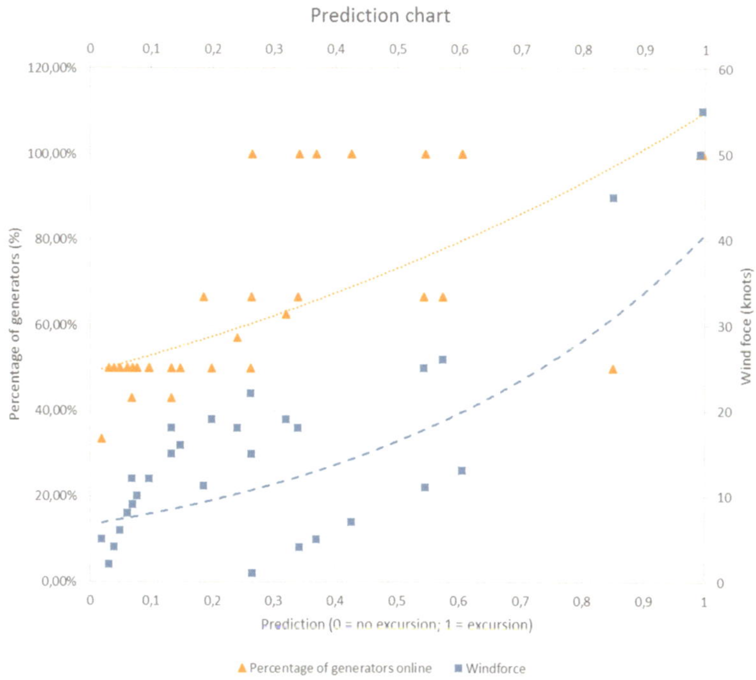

Figure 4.
Prediction chart showing the trends for wind force and percentage of generators according to the prediction model, for cases with no human cause.

position, four are correctly classified (40% of the total). There are 23 + 4 = 27 cases out of 33 that are correctly classified, representing 66.7% of the studied incidents.

In Step 2, when the variable percentage of generators is included in the equation, for the incidents not having excursion, the number of correctly-classified cases has become 20, representing 87% of the total. There is an improvement in the number of correctly classified cases for the excursion cases, which are now 5 (50% of the total). In this second step, there are now 25 cases correctly classified, which means 75.8% of the studied incidents. Although, there has been an evident downgrade in the prediction of the model with the addition of the variable percentage of generators, the prediction for the cases with loss of position has improved.

In **Figure 4**, it can be seen how the model predicts a loss of position for more significant values of wind force and of the percentage of generators.

The relative ratio can show the prediction for loss of position for the different main causes, as shown in **Figure 5**. The dashed line allows us to appreciate better those mean values above 1, which show a higher likelihood of having a loss of position. The main causes that are more prone to end in an excursion, according to this new model, are environmental and references.

5.3.2 Human cause

The 9 cases where there is no human cause are selected.

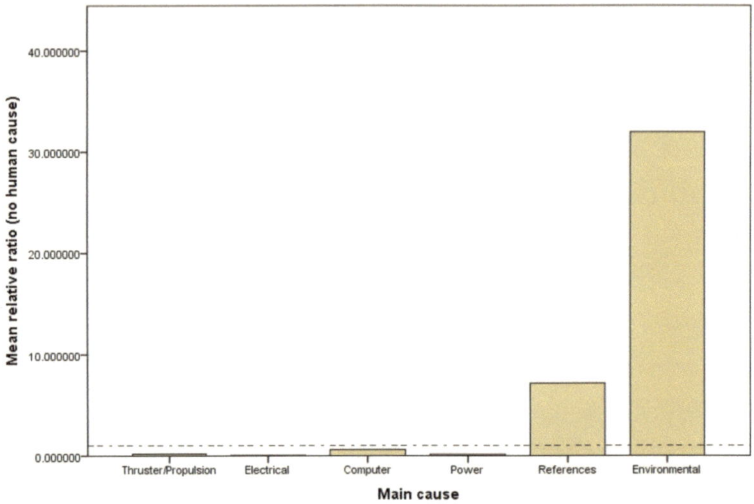

Figure 5.
Mean relative ratio for each main cause group, for the model obtained for the incidents without human cause, obtaining the likelihood of a loss of position.

Causal factor	n	B	Wald	p-value	Odds Ratio (Exp(B))	IC 95%	
						lower	upper
Water depth	9	−0.006	0.912	0.34	0.994	0.981	1.007
Percentage of thrusters	9	Not in the equation					
Percentage of generators	9	1.341	0	0.999	3.824	0	—
DGNSS	9	Not in the equation					
HPR	9	—	0	1	—	—	—
Taut wire	9	Not in the equation					
Inertia system	9	Constant value					
Gyros	9	Constant value					
MRUs	9	Constant value					
Wind sensors	9	Not in the equation					
Windforce	9	Not in the equation					
Force Beaufort	9	Not in the equation					
Wind direction	9	Not in the equation					
Current speed	9	Not in the equation					
Current direction	9	Not in the equation					
Wave height	9	Not in the equation					
Visibility ordinal	9	Not in the equation					

Table 9.
Individual results in step 1 for each independent variable when the forward (Wald) binary regression model is performed, being excursion the dependent variable and selection variable human cause = 1 (human cause).

Variables in the equation	B	S.E.	Wald	df	Sig.	Exp(B)	95% C.I.for EXP(B)	
							Lower	Upper
Waterdepth	−0.006	0.007	0.912	1	0.340	0.994	0.981	1.007
Constant	5.592	6.426	0.757	1	0.384	268.187		

Table 10.
Variables in the equation in Step 1. It can be observed that the independent variable, with a p-value of 0.34, cannot be considered to have any relation with the loss of position or not.

In the preliminary stage, the variables are introduced in the model one by one to check their significance for explaining the answer. The variables are presented in **Table 9**.

Considering these results, only the variable water depth is selected to enter the model.

In step 0, when the variable is not yet in the equation, it is considered significative (score 5.248, p-value 0.022), so it enters the equation in step 1. The different statistics obtained in Step 1 can be observed in **Table 10**.

It can be observed that the p-value associated with the Wald statistic is bigger than 0.1, which means that this variable does not explain the model with the desired significance, and so it must be rejected.

Out of the nine valid cases, six were not ending in an excursion, while three were losing position.

In Step 1, after the variable water-depth was included in the equation, it was obtained that from the 6 cases without excursion, according to the model, there were 5 cases correctly classified (83.3% of the total) and that from the ones having a loss of position, two are correctly classified (66.7% of the total). There are $5 + 2 = 7$ cases out of 9 that are correctly classified, representing 77.8% of the studied incidents.

6. Discussion

The first approximation to the regression model was to include the variables one by one to determine which variables could explain the answer.

It was interesting that the categorical variables did not explain the model. However, the data analysis could suggest that, had the sample been more prominent, they could have influenced the result. This study aimed not to distort the sample by performing a bootstrapping, not only because of the possible distortion of the sample but also because of the complication implied when the variables do not follow a normal distribution. The sample size is considered to be representative of the period of study.

With the first approach, the variables that could explain the probability of an excursion are determined to be: water-depth, percentage of generators online, Wind force and Wave height. The first two variables belong to the DP system configuration, and the last two are related to meteorological conditions.

However, when all of them are entered into the model, we obtained that only the first two explain the answer, while the other two could not improve the model already created. Although, this could give an idea of the less importance of the meteorological variables when explaining the excursion, it should not be forgotten that the meteorology, and especially the wind force (which creates waves with a height that is proportionally correlated to the force in knots), can also influence the probability of a unit having a loss of position while performing DP drilling operations.

The two selected independent variables, water-depth and percentage of generators, can explain the probability of losing position. This possibility will increase when the water depth has small values and the percentage of generators has significant values. These results are very interesting from the operator's point of view, as the lower values of water depths have traditionally been a common drilling ground where DP was not necessary, and other methods were used to achieve the position keeping. This could partly explain the problems of DP station keeping incidents when the drilling operations take place in shallow waters.

Studying the mean relative ratios for each main cause group, it is interesting to note that the model can explain environmental-, computer- and human-caused incidents more precisely than other causes. Within the group of human causes, the incidents without a human cause have a better prediction using the proposed model.

In general, this model correctly classifies 79% of the incidents, which is considered to be a very good prediction overall.

When the data is split into subgroups defined by the existence or not of a human cause, it can be seen how the mean percentage of thrusters is significantly more prominent for the cases where there is no human cause and smaller when there is a human cause. However, it does not explain whether the probability of an excursion is bigger or smaller for any of the subgroups. However, it can suggest that it would be a significant variable when the dependent variable is used to determine the possibility of a human error.

Always taking the general regression model from above into account, the regression model for the subgroup without human cause proposes the percentage of generators and wind force as variables that could explain the model. Wave height is not significant (although with a p-value of 0.104, it could be said that it is at the edge of being significant), and the water depth does not even enter the equation.

The fact that water depth was not even entering the equation when considered individually suggests it does not influence the probability of excursion when the cause is not human.

When studying the cases in the subgroup human cause, we obtain that the only variable that could explain the model is water depth. However, its p-value is bigger than 0.1, after entering the equation and because of this, it is rejected. In the iteration of the model, it can be seen how when this variable is included, the percentage of cases that are correctly classified decreases for the first group (no excursion). At the same time, the percentage of correctly-classified cases improves for group 2 (excursion). Probably the comparatively small number of cases (only 9 out of the 42 cases in total) contribute to the decision of rejecting this variable from the equation. Nonetheless, it suggests that this variable can be expected to be added to the model when a bigger sample is studied.

7. Conclusions

The purpose of this chapter was to determine the mathematical expression that explains the possibility of a loss of position during DP drilling operations.

With a sample of 42 incidents from 2011 till 2015, it was determined that the mathematical expression for the binary logistic regression model is shown in Eq. 9.

The loss of position of an incident depends on the water depth and the percentage of generators used.

With this model, it can be determined that the probability of loss of position will increase when the water-depth has small values and the percentage of generators has bigger values.

Having considered that the percentage of cases correctly classified by the model which takes into account both variables percentage of generators online and water-depth is high (78.6%), it is expected to provide excellent results when predicting whether any incident will have a loss of position or not.

Once this model has been determined, the secondary objective of this paper was to find and compare the mathematical expression, taking into account whether the nature of the cause leading to the incident was human or not.

With a sample of 33 incidents without human cause, it was determined that the mathematical expression for determining the probability of loss of position is given in Eq. 13. This model can determine that the probability of having an excursion increases as the percentage of generators and the wind force have bigger values.

The percentage of cases correctly classified by the model according to this sample is high (75.8%), and it can be expected to provide excellent results when predicting whether an incident that has no human origin will have a loss of position or not.

The sample of the incidents with human cause was relatively small (9 cases only), so no independent variable could explain the model within a confidence interval of 10%. However, the variable water-depth, which appears above in the general model, and does not appear in the model for the cases without human cause, can be suspected to explain the model, although it will be necessary to perform further research on a bigger sample to obtain significant results.

Conflict of interest

"The authors declare no conflict of interest."

Nomenclature

-2LL	-2LogLikelihood, statistic used to define the goodness of fit of a model
B	regression coefficients of the independent variables in the model
DGNSS	Differencial Global Navigation Satellite System
DP	dynamic positioning
DPO	dynamic positioning operator
E	Error between the estimated probability and the observed probability
HMI	Human Machine Interface
HPR	Hydro-acoustic Positioning References
IMCA	International Marine Contractors Association
LMRP	lower marine riser package
MRU	motion reference units
p	probability that the dependent variable will obtain a value 1
p/q	Mean relative ratio of the probability of the variable to obtain a value of 1
POUT	output value
PRS	position reference systems
q	probability that the dependent variable will obtain a value 0

SE	standard error
Wald	Statistic used to determine whether a variable is contributing to defining the model or not
Z	formula defining the model

Author details

Zaloa Sanchez-Varela
Faculty of Maritime Studies, University of Split, Split, Croatia

*Address all correspondence to: zsanchezv@pfst.hr

References

[1] Bray D. DP Operator's Handbook. 2nd ed. London: The Nautical Institute; 2015. ISBN: 9781906915254

[2] Bray D. DP Operations. Part 2 - What is DP used for? London: The Nautical Institute, Seaways; 2018. pp. 13-16

[3] Sørensen AJ. A survey of dynamic positioning control systems. Annual Reviews in Control. 2011;**35**(1): 123-136. DOI: 10.1016/j.arcontrol.2011.03.008

[4] Mao L, Zeng S, Liu Q. Dynamic mechanical behavior analysis of deep water drilling riser under hard hang-off evacuation conditions. Ocean Engineering. 2019;**183**:318-331. DOI: 10.1016/j.oceaneng.2019.05.014

[5] Hogenboom S, Rokseth B, Vinnem JE, Utne IB. Human reliability and the impact of control function allocation in the design of dynamic positioning systems. Reliability Engineering & System Safety. 2020;**194**:106340. DOI: 10.1016/j.ress.2018.12.019

[6] The Nautical Institute. Accreditation and Certification Scheme Standard. London: The Nautical Institute; 2022

[7] Chen H, Moan T, Verhoeven H. Safety of dynamic positioning operations on mobile offshore drilling units. Reliability Engineering & System Safety. 2008;**93**(7):1072-1090. DOI: 10.1016/j.ress.2007.04.003

[8] Verhoeven H, Chen H, Moan T. Safety of dynamic positioning operation on mobile offshore drilling units. Houston: MTS Dynamic Positioning Conference; 2006

[9] Chae C. A Study on Human Error of DP Vessels LOP Incidents. Journal of the Korean Society of Marine Environment & Safety. 2015;**21**(5):515-523. DOI: 10.7837/kosomes.2015.21.5.515

[10] Chae C. A Study on FSA Application for Human Errors of Dynamic Positioning Vessels Incidents. Journal of Korean Navigation and Port Research. 2017;**41**(5):259-268. DOI: 10.5394/KINPR.2017.41.5.259

[11] Dong Y, Vinnem JE, Utne IB. Improving safety of DP operations: Learning from accidents and incidents during offshore loading operations. EURO Journal on Decision Processes. 2017;**5**(1–4):5-40. DOI: 10.1007/s40070-017-0072-1

[12] Øvergård KI, Sorensen LJ, Nazir S, Martinsen TJ. Critical incidents during dynamic positioning: Operators' situation awareness and decision-making in maritime operations. Theoretical Issues in Ergonomics Science. 2015;**16**(4):366-387. DOI: 10.1080/1463922X.2014.1001007

[13] Marine Forecasts Glossary. 2019. Available from: https://www.metoffice.gov.uk/weather/guides/coast-and-sea/glossary. [Accessed: 2022-01-09]

Chapter 11

On the Use of Modified Winsorization with Graphical Diagnostic for Obtaining a Statistically Optimal Classification Accuracy in Predictive Discriminant Analysis

Augustine Iduseri

Abstract

In predictive discriminant analysis (PDA), the classification accuracy is only statistically optimal if each group sample is normally distributed with different group means, and each predictor variance is similar between the groups. This can be achieved by accounting for homogeneity of variances between the groups using the modified winsorization with graphical diagnostic (MW-GD) method. The MW-GD method involves the identification and removal of legitimate contaminants in a training sample with the aim of obtaining a true optimal training sample that can be used to build a predictive discriminant function (PDF) that will yield a statistically optimal classification accuracy. However, the use of this method is yet to receive significant attention in PDA. An alternative statistical interpretation of the graphical diagnostic information associated with the method when confronted with the challenge of differentiating between a variable shape in the groups of the 2-D area plot remains a problem to be resolved. Therefore, this paper provides a more comprehensive analysis of the idea and concept of the MW-GD method, as well as proposed an alternative statistical interpretation of the informative graphical diagnostic associated with the method when confronted with the challenge of differentiating between a variable shape in the groups of the 2-D area plot.

Keywords: winsorization, informative graphical diagnostic, optimal training sample, predictive discriminant analysis, statistically optimal classification accuracy

1. Introduction

Binary classification when compared to multiple classification has wide range of real-world applications in many areas of human endeavors, such as criminal justice, education, medicine, email analysis, human resources management, pattern recognition, energy and environmental management, financial data analysis and economics, production systems management and technical diagnosis, marketing among others. Where the classification criterion comprises one or several predictor variables along with a categorical criterion, such a prediction will require the use of a predictive discriminant analysis (PDA). PDA is still the optimal method when the cost of misclassifying groups is clearly different and when there is greater interest in the accuracy of classifying separate groups. In most cases, evaluating the proportion of correct classification of a predictive discriminant function (PDF) in all subpopulations is equivalent to the estimation of the actual hit rate, $P^{(a)}$ [1, 2]. That is, $P^{(a)}$ is the expected proportion of correct classification when a PDF that is built from a given training sample is validated on training samples from the sample population. In PDA, to improve or optimize classification accuracy or actual hit rate, researchers often rely on feature selection methods. The aim of feature selection methods in PDA is to choose the best subset of important predictor variables that will effectively reduce the intricacy of the PDF, thus facilitate interpretation, enhance or optimize the classification accuracy, and reduce the training time. Nevertheless, the promise of optimizing classification accuracy using variable selection methods is almost always unfulfilled, because the derived PDF is often obtained from a training sample that does not meet near optimal condition [1, 3–6]. The actual hit rate of a PDF may be considered statistically optimal only if the assumptions of normality and/or homogeneity of variances are taken into account [5, 7]. This means that having a better subset is not a guarantee for achieving a statistically optimal classification accuracy.

In general, the task of enhancing or improving classification accuracy was examined in two ways. Several researchers use feature or variable selection techniques to select the best subset of predictors to construct a classification model. In addition to conventional feature selection techniques, including the stepwise and all possible subset methods [4, 8, 9]. Some widely known and used methods include the principal component analysis (PCA) used to obtain a set of low-dimensional features from a large set of features [10, 11]. The branch and bound technique which uses a greedy procedure to obtain the best subset [12], the genetic search algorithm [13, 14], the shrinkage methods [10, 15], the particle swarm optimization (PSO) approach which is a meta-heuristic technique used to enhance classification accuracy [16], representative methods based on dictionary learning (DL) for classification [17–19], support vector machines (SVMS) [20], and the hyper parameter tuning approach [21, 22]. We have the sequential analysis approach as well [23]. The heteroscedastic discriminant analysis merged with feature selection [24] and the modified leave-one-out (LOOCV) cross-validation method used as an alternative to the all-possible subset method [25]. A PDF's classification accuracy is only statistically optimal if each group sample is normally distributed with different group means, and each predictor variance is similar between the groups [7]. None of these basic assumptions regarding the validity/reliability of the PDF are considered by any of the above methods. To address these gaps in feature selection techniques, other investigators are seeking alternatives to robust PDA by replacing conventional estimators with robust estimators. Some variants of these alternative methods include the dimensionality reduction/feature extraction for outlier detection (DROUT) [26], the minimum covariance determinant (MCD) [27], S-estimators [28], one-step M-estimator (MOM), and winsorized one-step M-estimator (WMOM) [29]. These other methods concentrate on building a robust PDF for deviations due to the presence of outliers in the training sample. Besides the presence of outliers in most training samples, there are also hidden influential observations resulting either from an incorrect distributional assumption or an inherent variability of the dataset [30]. Oftentimes, these hidden influential observations are not considered by any of the above methods for optimizing an actual hit rate. Consequently, the PDF's solution obtained by either of the two approaches may be optimal but not statistically optimal. To overcome the problem of hidden influential observations, Iduseri and Osemwenkhae [6] proposed a novel method for attaining an optimal training sample. Their method otherwise known as modified winsorization with graphical diagnostic (MW-GD) method yielded a PDF's solution which was statistically optimal for both the training sample that gave rise to it and for other training samples from the same population. However, the graphical diagnostic associated with this new method may be difficult to interpret if there are no significant differences between a variable shape in the groups of the 2-D area plot, and yet there is evidence of hidden influential observations in the training sample.

This paper provides a more comprehensive analysis of the idea and concept of the MW-GD method, as well as proposed an alternative statistical interpretation of the informative graphical diagnostic associated with the method when confronted with the challenge of differentiating between a variable shape in the groups of the 2-D area plot. The remaining sections of this paper are organized as follows. Sections 2 and 3 discuss the problems posed by the presence of outliers and legitimate contaminants in the training sample that yields the PDF, the concept of statistical optimality of the PDF classification accuracy, and the robustness of PDF, respectively. Section 4 describes in details the idea and concept of the modified winsorization with a graphical diagnostic for obtaining a statistically optimal training sample, as well as presents the proposed alternative statistical or numerical interpretation of the informative graphical diagnostic. Section 5 presents the results and discussions based on the application of two real life samples, while Section 6 presents the conclusions.

2. Outliers and legitimate contaminants in PDA

In PDA, an outlier is an observation which is not a member of a group, and is often indicative of an incorrect measurement or an incorrect allocation of the unit or observation. Such an outlying observation can cause severe problems that even the robustness of PDA may not overcome. Over the last two decades, many articles have been published about detecting outliers in discriminant analysis (DA) [31–38]. In PDA, a popular means of treating outliers is to construct multiple PDFs with assumed outliers added and with assumed outliers removed [1]. The primary issue with this method is whether potential outliers should be remove one at a time, two at a time, or all at a time. With the SPSS DISCRIMINANT procedure, the chi-squared distribution is used to establish the typicality probability. These typicality probabilities are used to identify potential outliers in the context of PDA. However, Huberty and Olejnik [1] pointed out that when the group covariance matrices are not equal, the unit typicality probabilities are difficult to interpret because different distance metrics are used in the calculation. A common distance index used for detecting outliers or influential observations in the context of PDA is the Mahalanobis distance [39] which is also calculated as a byproduct in SPSS DISCRIMINANT procedure.

However, there are also hidden influential observations (or legitimate contaminants) resulting either from an incorrect distributional assumption (i.e., when the data turns out with a different structure than originally assumed) or an inherent variability of the dataset, see Osborne [30], and Iglewicz and Hoaglin [40] for more details. While hidden influential observations may actually belong to a training sample, but if not distributed randomly may reduce normality which often leads to violation of sphericity and multivariate normality assumptions in PDA. Hidden influential observation can also adversely affect the quality of the PDA solution and its generality. But how to identify and remove hidden influential observations before building a classification model (particularly in the PDA) has not receive any significant attention in the literature by statisticians or by methodologist and therefore not by any substantive researchers. Besides, the SPSS typicality or Mahalanobis index may not be able to identify hidden influential observations because their unit often belongs to a different group compared to outliers.

Therefore, much emphasis should be placed on cleaning the training sample to ensure that it meets its near-optimum condition by removing all legitimate contaminants from the training sample. This method is similar to optimizing decision trees (in particular classification trees) which consists in reducing the amount of impurity—see Myatt [41] for details. In the context of PDA, it will improve the similarity of each predictor variable variance between groups, thus improving the approximation of the true shape. This will in no doubt guarantee the statistical optimality of the PDF solution or classification accuracy.

3. Statistical optimality of a PDF classification accuracy

A PDF's classification accuracy is only statistically optimal if each group sample is normally distributed with different group means, and each predictor variance is similar between the groups [5, 7]. In addition to the above requirements, it is recommended that there be at least four to five times as many cases as predictors in order to produce more accurate estimates. Note that in PDA, the failure of the training sample to meet the assumption of normality can result in a decrease in efficiency and accuracy—see Lachenbruch [42] as cited in Klecka [43]. However, a minor violation of this assumption will not decrease the accuracy of the classification. As long as the distributions of predictors are reasonably comparable, the estimation of most multivariate parameters does not require multivariate normality [44]. Moreover, under the central limit theorem, there is no need to worry about the assumption of normality as long as each group sample contains a very large number of observations. As a general rule, a PDF will still perform strongly against non-normality as long as the smallest group has over 20 cases, and the number of predictors is less than six [45]. Due to these robustness properties of PDA, researchers are barely concerned about the assumption of normality.

But where non-normality is due to outliers and/or hidden influencing observations other than skewness, violating this assumption has serious consequences, because PDA is very sensitive to outliers [45]. Likewise, more cases could be classified into the group with greater dispersion when the assumption of equality of variance-covariance matrices is not tenable [45]. In addition, the likelihood of belonging to a group may be distorted, and the PDF may also not be able to separate the groups as much as possible [43]. The accuracy of the discriminant weights estimates may be reduced if the variances of the predictors are not all similar between groups. They may be precise but not unbiased [46]. When the homogeneity of variance test is significant, it indicates that the training sample is contaminated with outliers and/or hidden influence observations, and the significance tests are unreliable [3, 45]. It is apparent from the foregoing that, if the assumption of homogeneity of variances is not satisfied, it is probable that the assumption of multivariate normality is not equally satisfied. This suggests that multivariate normality and homogeneity of variance assumptions can be taken into account if outliers and hidden influential observations are completely removed from a training sample. The practice of researchers relying on the robustness properties of PDA without checking for outliers and hidden influential observations, which may hinder maximal separation between the groups seems to be a norm. This practice is further encouraged by the general acceptance of a hit rate of 25% above that of chance. Assuming that you get a 95% hit rate, it is certain that you will not care about the two basic assumptions of PDA.

Whereas the reason for such good performance might be that the data support simple linear or quadratic separation boundaries. The general belief that linear classifiers are robust to minor violations of its basic assumptions (in particular is the assumption of multivariate normality) is often not tenable. Studies have shown that the reliability of a PDF solution is dependent upon adherence to the underlying assumptions [5]. The primary objective of PDA is classification, and if the percentage of correct classifications is not satisfactory, it is likely that variances in predictors are not similar across groups. That is to say the training sample is not statistically optimal. Therefore, it is necessary to adopt a screening method that will effectively identify and remove legitimate contaminants from training samples before using them to build a PDF. Iduseri and Osemwenkhae [6] proposed the modified winsorization with graphical diagnostic (MW-GD) method to identify and remove legitimate contaminants from training samples. The MW-GD method produced a statistically optimal training sample when applied to a real dataset, and the resulting PDF yielded a hit rate that was statistically optimal. As a result, the uncertainty about the PDF's actual hit rate was greatly reduced. However, the informative graphical diagnostic associated the proposed method may be difficult to interpret if there are no significant differences between a variable shape in the groups of the 2-D area plot.

This paper proposes an alternative statistical interpretation of the informative graphical diagnostic when confronted with the challenge of differentiating between a variable shape in the groups of the 2-D area plot.

4. Identification and removal of legitimate contaminants in PDA

4.1 Correction for bias of discriminant weights in PDA

In general, extreme scores or outliers bias estimates of any parameter. One notable way to correct for bias is to change the data by changing the scores so as to reduce the impact of the extreme scores or adjust the shape of the distribution. Notable variants of changing the scores method include transforming the data, trimming the data and winsorizing the data. However, in PDA, where one is interested in differences between set of variables or groups, transformation may not be a good choice to correct for bias of discriminant weights. This is because, transformation can change the units of measurements, which may in turn affects the interpretation of the data because the data now relate to a different construct compared to the original data [47]. Similarly, trimming of data seemed odd since one could just discard lots of data. To overcome these inherent drawbacks associated with both methods, the winsorization method was adopted. These approach involves replacing a percentage of the highest score with the next highest score in the data and the same percentage of the lowest score are replaced with the next lowest score in the data. One major challenge with this method is that even the next higher or lowest score might still be an extreme score. Another variant of the winsorization involves replacing extreme score with a score three standard deviations from the mean. This variant of winsorization also suffers a major drawback. As noted by Field [47], the standard deviation will be biased by extreme scores, so this means that you are replacing scores with a value that has been biased by extreme scores. To address the observed shortcomings of both variants of the winsorization method, Iduseri and Osemwenkhae [6] proposed the modified winsorization with graphical diagnostic (MW-GD) method. The method proved very effective in identifying and removing legitimate contaminants.

4.2 The modified winsorization with graphical diagnostic (MW-GD) method

In this section, the modified winsorization with graphical diagnostic (MW-GD) method originally proposed by Iduseri and Osemwenkhae [6] is presented. In addition, a proposed alternative statistical interpretation of the informative graphical diagnostic associated with MW-GD method when confronted with the challenge of differentiating between bar shapes of the 2-D area plot is also presented. The MW-GD method, which involves a three-step procedure, will effectively identify and eliminate legitimate contaminants from predictor variables so that their variances between the groups are similar. The aim is to ensure that the training sample, D_N^t satisfies the basic assumptions (particularly the assumption of homogeneity of variances) of the PDA. The three steps procedure produced an optimal training sample that was used to construct a PDF whose percentage of correct classification was not only statistically optimal for the training sample that produced it, but also for other training samples from the same population.

4.2.1 Algorithm for the modified winsorization with graphical diagnostic (MW-GD)

Let $D_N^t = [x_1, x_2, \cdots, x_N] \in \Re^{P \times N}$ be a training sample matrix that comprises N observations set$\{(x_i, y_i)\}_{i=1}^{n}$, obtained from a real dataset using any of the conventional feature selection techniques, where $x_i \in \{1, 2, \cdots, P\}$ denote the corresponding predictor variable label, $y_i \in \{1, 2, \cdots, K\}$ denote the corresponding group label, P is the number of predictor variables, and K is the number of groups.

Step 1: Identification of the predictor variables with legitimate contaminants.

For the training sample, D_N^t the scores or observations of the predictor variables, $X_1, \cdots, X_N$ are first arranged in ascending order so that the extreme scores at both ends can be identified. For each predictor variable in the unordered training sample, D_N^t, a pair of score (i.e., the most extreme and the least extreme scores initially identified at both ends of each distribution) is deleted and replaced with the median value, before the mean of the remaining scores is calculated. The median value was adopted in order to satisfy the assumption of independence of all cases. The median is a position average independent from all other cases, whereas the mean depends on all other cases. Consequently, substituting the identified influential observations with their median value takes into account the assumption that all observations must be independent. This process is repeated for other lower pairs of extreme values and stops when five modified winsorized means are obtained for each predictor variable. When the modified winsorized means values are plotted, the predictor variable with bar shapes that are not similar between groups in the 2-D area plot becomes the predictor variable with legitimate contaminants.

Step 2: Removing legitimate contaminants from the identified predictor variables.

To determine what percentage of winsorization is required to eliminate the legitimate contaminants, the modified winsorization process is repeated only for the predictive variable(s) identified with the legitimate contaminants until the highest hit rate is attained, thus obtaining a near optimal training sample given as:

$$D^t_{N(optimal)} = [x_1, x_2, \cdots, x_P] \tag{1}$$

Step 3: Obtaining a statistically optimal hit rate.

The optimal training sample of Eq. (1), was then used to build the optimized PDF, $Z_{(opt)}$ given as:

$$\begin{aligned} Z_{(optimal)} &= u_1X_1 + u_2X_2 + \cdots + u_PX_P \\ &= \eta\left(D^t_{N(optimal)}\right) \end{aligned} \tag{2}$$

where $Z_{(optimal)}$ is the optimized PDF, u_i are the discriminant weights, X_i are the predictor variables and $\eta\left(D^t_{N(optimal)}\right)$ shows that the PDF is constructed with an optimal training sample.

To get a statistically optimal hit rate, let:

$$d_j = \begin{cases} 1 & if\, \hat{Z}_j = Z_j \\ 0 & otherwise \end{cases} \tag{3}$$

where $\hat{Z}_j$ is the predicted response for the *jth* observation in the optimized training sample, Z_j is the value for the *jth* observation in the optimized training sample. Therefore, a statistically optimal hit rate for the optimized PDF in (2) is given as:

$$P^{(a)} = \frac{1}{N^t}\left(\sum_{j=1}^{N^t} d_j\right) \times 100 \tag{4}$$

where N^t is the total number of cases over all groups in the optimized training sample. If we redefine d_j as:

$$d_j = \begin{cases} 1 & if\, \hat{Z}^{-j} = Z_j \\ 0 & otherwise \end{cases} \tag{5}$$

where $\hat{Z}^{-j}$ is the predicted response for the *jth* observation computed with the *jth* observation removed from the training sample. The leave-one-out cross-validation (LOOCV) estimate of the optimized hit rate (4) is given by:

$$\hat{P}^{(a)}_{LOOCV} = \frac{1}{N}\left(\sum_{j=1}^{N_v} d_j\right) \times 100 \tag{6}$$

where n_v is the number of validation samples.

4.3 The proposed alternative statistical interpretation for the informative graphical diagnostic

As described in Step 1 of Section 4.2.1, the graphical representation of the modified winsorized means makes it possible to easily identify the predictor variable(s) whose variance is not significantly similar between the groups. If the variance of a predictor variable is not similar between groups, the bars representing the modified winsorized mean values for the variable in the 2-D area plot will not have similar shape. Such a variable is interpreted to be the identified variable that contains legitimate contaminants. This means that the more the shape of the bars differ between the groups, the easier it is to interpret the 2 D area plot. It is therefore necessary to provide an alternative interpretation when it is difficult to differentiate between a variable shape in the groups of the 2-D area plot. Therefore, a simple statistical or numerical interpretation is proposed for the informative graphical diagnosis when it is difficult to differentiate between a variable shape in the groups of the 2-D area plot. This alternative numerical interpretation consists of the following two simple steps:

Step 1: Fitting the modified winsorized means values to a linear regression model.

For each group, the modified winsorized means and their corresponding winsorized percentage values for each predictor variable are fitted to a linear regression model given as:

$$\begin{aligned} Y_{11} &= a + b_{11}X \\ &\vdots \\ Y_{1P} &= a + b_{1P}X \end{aligned} \tag{7}$$

and

$$\begin{aligned} Y_{21} &= a + b_{21}X \\ &\vdots \\ Y_{2P} &= a + b_{2P}X \end{aligned} \tag{8}$$

where, $Y_{11}, \cdots, Y_{1P}$ and $Y_{21}, \cdots, Y_{2P}$ are the modified winsorized mean values for the P predictor variables in groups 1 and 2, respectively, and X is the corresponding winsorized percent values.

Step 2: Obtaining the absolute difference between the corresponding regression coefficients for the groups.

The absolute difference between the obtained regression coefficients (i.e., the slope) in group 1 and 2 is computed as:

$$\delta_{abs} = b_{1i} - b_{2i}, i = 1, \cdots, p \tag{9}$$

where the subscripts 1 and 2 represents groups 1 and 2. The predictor variable that has an absolute difference of 0.75 or greater will be the variable identified with legitimate contaminants. In PDA, if two samples are equal in size, there is always a 50/50 chance. Most researchers would accept a classification accuracy of 25% greater than that caused by chance. Hence, the choice of 0.75 as the decision boundary.

5. Computational results and discussion

5.1 Using the modified winsorization with graphical diagnostic (MW-GD) method

To evaluate the effectiveness of the modified winsorization with graphical diagnostic (MW-GD) method, two real samples (see [6] for the two data sets) were considered with the second used as a validation sample. The first training sample is from a renowned financial journal, among Japanese business leaders, which can be compared to the Economist, Financial Times, and Business Week in Europe and the United States of America. This dataset contains 50 observations from each of the two groups of Japanese financial institutions, each bank being evaluated using the following seven performance indexes: (1) return on total assets (= total profits/average total assets), (2) labor profitability (= total profits/total employees), (3) equity to total assets (= total equity/average total assets), (4) total net working capital, (5) return on equity (= earnings available for common/average equity), (6) cost-profit ratio (= total operating expenditures/total profits), and (7) bad loan ratio (= total bad loans/total loans). However, taking into account the beneficial effect of feature selection and outlier detection as a preprocessing step, a best subset and critical value of Mahalanobis distance were first obtained using the SPSS stepwise method, and its compute command for critical value of Mahalanobis distance. The stepwise approach produced a best subset which comprises return on total assets (X_1), labor profitability (X_2), equity to total assets (X_3), and bad loan ratio (X_7). The SPSS output of the constructed PDF based on the training sample of four variables given by:

$$Z = 0.005X_1 + 0.006X_2 + 0.004X_3 + 0.005X_7 \tag{10}$$

Thus, the classification accuracy of the PDF, Z (10) and its LOOCV estimate are given by Eqs. (11) and (12).

$$P^{(a)} = \frac{1}{N^t}\left(\sum_{j=1}^{N^t} d_j\right) \times 100 = 86.0\% \tag{11}$$

$$\hat{P}^{(a)}_{LOOCV} = \frac{1}{N}\left(\sum_{j=1}^{n_v} d_j\right) \times 100 = 85\% \tag{12}$$

While the two critical values from the SPSS outputs for 0.95 and 0.99 used as the probabilities in the IDF.CHISQ function with four predictor variables were given as:

$$COMPUTEcritical = IDF.CHISQ\ (0.95, 4) = 9.49$$
$$COMPUTEcritical = IDF.CHISQ\ (0.99, 4) = 13.28$$

The Mahalanobis distance values for all cases, as reported in the case wise statistics table, were all lower than the two critical values. This means that there are neither outliers nor hidden influential observations in the dataset or training sample. Next, the MW-GD algorithm was applied to the training sample, D^t_N consisting of four predictor variables.

At *step 1* of the MW-GD method, a total of five modified winsorized means for the four predictor variables were obtained. The summary of the winsorized means values is presented in **Table 1**.

To provide a meaningful interpretation of **Table 1**, the modified winsorized means or averages for both groups, as shown in **Table 1**, were plotted with a 2D area plot in Excel Package. The process involves entering the first variable X_1 modified winsorized averages for both groups into Excel spreadsheet in pairs (with X_1 values for group 1 and 2 occupying the first 2 columns from row 1 to row 6), followed by variable X_2 (with X_2 values for group 1 and 2 occupying column 3 to column 4 from row 8 to 13), followed by variables X_3 and X_7 proceeding downward in steps. The graphical representation is presented in **Figure 1**.

Groups	No. of replaced sample points	Modified winsorized mean values				Winsorized %
		X_1	X_2	X_3	X_7	
1	0	428.70	329.48	536.34	821.32	0
	2	421.08	317.32	533.90	824.50	4
	4	415.50	307.60	533.70	827.40	8
	6	412.28	298.32	533.56	830.36	12
	8	408.90	290.30	534.46	833.62	16
	10	406.20	282.70	535.42	835.56	20
2	0	341.50	177.20	368.46	778.70	0
	2	335.24	173.20	366.92	785.80	4
	4	329.66	169.16	365.20	790.74	8
	6	324.76	166.68	362.94	794.22	12
	8	319.86	166.22	360.68	797.30	16
	10	315.44	165.94	358.66	800.12	20

Table 1.
Modified winsorized means for up to five pairs of winsorized values.

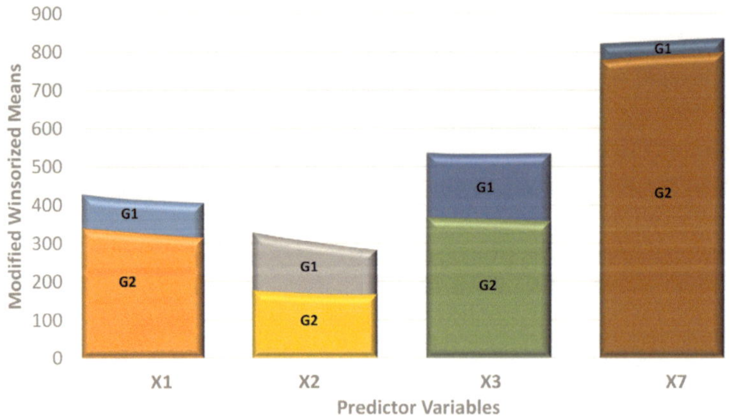

Figure 1.
Graphical representation of the modified winsorized mean values in ***Table 1****.*

A cursory look at **Figure 1** shows that the winsorized average values for the four predictor variables in both groups represented by the 2-D area plot have similar shape (or have similar variances within the groups) except for predictor variable X_2. Also, in **Figure 1**, the bar shape of the predictor variable X_2 in group 1 looks like a rectangle whereas in group 2 it looks like a trapezium. The observed difference in the shape of the variable X_2 bar indicates that variable X_2 does not have similar variances in the groups, and therefore becomes the only variable with legitimate contaminants. It therefore means that the training sample does not satisfy the assumption of homogeneity of variances. This finding was corroborated by the result of the Box M test for equality of variance-covariance matrices for this training sample, which was significant. Apart from manually replacing the extreme values on both ends with the median value, for each percentage of winsorization using R aggregate () function, the average calculation time needed to generate each row result for groups 1 and 2 in **Table 1** was 2 seconds.

At *step 2* of the MW-GD method, the modified winsorization process was performed for only variable X_2. For each percent of winsorization, a PDF is constructed using all four predictor variables. The summary of hit rate results for each percent of winsorization is presented in **Table 2**. **Table 2** shows that the highest hit rate of 97.00 was achieved when 5 data points at both ends of the data were replaced by the median value. This means that all the legitimate contaminants in variable X_2 was completely replaced at 20% winsorization. In other words, at 20% winsorization, the lack of homogeneity of variances observed in the variable X_2 was taken into account, thus obtaining a near optimal training sample, $D_{N\,(opt)}$:

The optimized training sample, $D_{N\,(opt)}$ was then used to construct a PDF. The SPSS output for the obtained PDF is given as:

$$Z_{opt} = 0.003X_1 + 0.018X_2 + 0.001X_3 + 0.004X_7 \tag{13}$$

Thus, the PDF, Z_{opt} (13) hit rate and its LOOCV estimate are given by Eqs. (14) and (15).

$$P^{(a)} = \frac{1}{N^t}\left(\sum_{j=1}^{N^t} d_j\right) \times 100 = 97.0\% \tag{14}$$

$$\hat{P}^{(a)}_{LOOCV} = \frac{1}{N}\left(\sum_{j=1}^{n_v} d_j\right) \times 100 = 91\% \tag{15}$$

% of winsorization	No. of replaced sample points	% of training sample correctly classified
4	2	85.00
8	4	88.00
12	6	91.00
16	8	94.00
20	10	97.00[a]
24	12	96.00

[a]*Optimal winsorization occurs at 20% with hit rate = 97.00%.*

Table 2.
Summary of hit rate results for each percent of modified winsorization for predictor variable, X_2.

In addition to the dataset from Japanese banks, a second real dataset was used to validate the first sets of results (11), (12), (14), and (15). This validation sample was obtained from the academic records of junior secondary school (JSS) 2, University Demonstration Secondary School (UDSS), University of Benin, Nigeria. The dataset contains 30 observations for both classes: Science and Art. The dataset consists of average scores for the three consecutive terms obtained for eleven (11) subjects, including English Language (X_1), Mathematics (X_2), Integrated Science (X_3), Social Studies (X_4), Introductory Technology (X_5), Business Studies (X_6), Home Economics (X_7), Agricultural Science (X_8), Fine Art (X_9), Physical and Health Education (X_{10}), and Computer Studies (X_{11}). Using the SPSS stepwise method, a subset of three variables comprising introductory technology (X_5), physical and health education (X_{10}), and computer science (X_{11}) was obtained. The SPSS output for the PDF is given as:

$$Z = 0.135X_5 - 0.102X_{10} + 0.058X_{11} \tag{16}$$

Thus, the PDF, Z (16) hit rate and its LOOCV estimate are given by Eqs. (17) and (18).

$$P^{(a)} = \frac{1}{N^t}\left(\sum_{j=1}^{N^t} d_j\right) \times 100 = 85.0\% \tag{17}$$

$$\hat{P}^{(a)}_{LOOCV} = \frac{1}{N}\left(\sum_{j=1}^{n_v} d_j\right) \times 100 = 83\% \tag{18}$$

Also, the two critical values from the SPSS outputs for 0.95 and 0.99 used as the probabilities in the IDF.CHISQ function with three predictor variables were given as:

$$COMPUTEcritical = IDF.CHISQ\ (0.95, 4) = 7.81$$
$$COMPUTEcritical = IDF.CHISQ\ (0.99, 4) = 11.34$$

The Mahalanobis distance values for all cases, as reported in the case wise statistics table, were all lower than the two critical values. This means that there are neither outliers nor hidden influential observations in the dataset or training sample. Once again, the proposed algorithm is applied to this second training sample, D^t_N consisting of three predictor variables.

At *step 1* of the MW-GD method, a total of five modified winsorized means for the three predictor variables were obtained. The summary of the modified winsorized means values is presented in **Table 3**.

Again, to interpret **Table 3**, the modified winsorized means or averages for both groups, as shown in column three of **Table 3**, were plotted with a 2D area plot in Excel Package. The graphical representation is presented in **Figure 2**.

A cursory look at **Figure 2** shows that the winsorized average values for the four predictor variables in both groups represented by the 2-D area plot have similar shape (or have similar variances within the groups). The similar shape shown by the three variables in each group indicates that there are no legitimate contaminants in the training sample, D^t_N. This implies that the fit between the training sample, D^t_N and the basic assumptions of PDA is sufficient (in particular is the assumption of homoscedasticity) to construct a PDF whose hit rate can be said to be statistically optimal. Therefore, for this dataset, the MW-GD algorithm ends at *step 1*. The initial training

Groups	No. of replaced sample points	Modified winsorized mean values			Winsorized %
		X_5	X_{10}	X_{11}	
1	0	67.27	69.77	73.30	0.00
	2	67.40	70.07	73.43	6.67
	4	67.43	70.17	73.60	13.33
	6	67.47	70.13	73.67	20.00
	8	67.53	70.13	73.70	26.67
	10	67.60	70.17	73.73	33.33
2	0	51.47	62.40	63.93	0.00
	2	51.27	62.53	63.97	6.67
	4	51.07	62.70	63.73	13.33
	6	50.97	62.83	63.60	20.00
	8	50.77	63.00	63.50	26.67
	10	50.73	63.23	63.30	33.33

Table 3.
Winsorized means for up to five pairs of winsorized values.

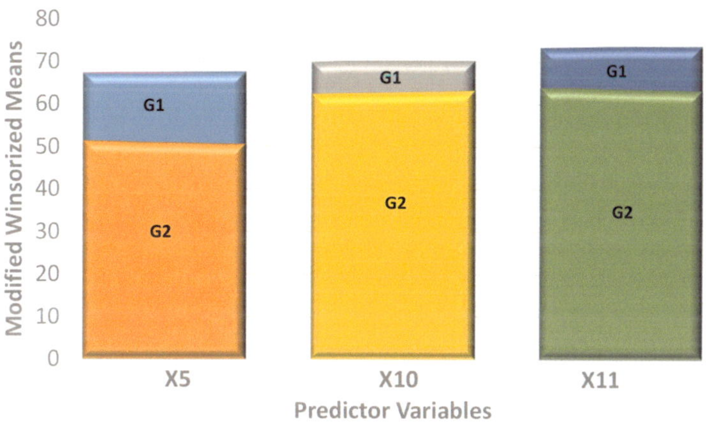

Figure 2.
Graphical representation of winsorized mean values in ***Table 3****.*

sample of three variables obtained from the second real dataset using SPSS stepwise method is therefore an optimal training sample.

5.2 Using the proposed alternative statistical interpretation for the informative graphical diagnostic

If the modified winsorized means for each variable in **Table 1** are denoted as the independent variable Y ($Y = Y_1, Y_2, Y_3, Y_4$), and the winsorized percent as the dependent variable X, then the six pairs of values for X (0, 4, 8, 12, 16, 20) and Y_1

Group	Regression coefficients, *b*			
	X_1	X_2	X_3	X_7
1	−1.088	−2.316	−0.022	0.725
2	−1.295	−0.569	−0.500	1.036
$\delta_{abs} = b_{1i} - b_{2i}$	0.2	1.8	0.5	0.3

Table 4.
Regression coefficients, b *of fitted regression model to* ***Table 1*** *data.*

(428.70, 421.07, 415.50, 412.28, 408.90, 406.20) are in good accordance to a regression model.

At *step 1* of the proposed alternative statistical interpretation for the informative graphical diagnostic, each of the six values of the modified winsorized means for the four variables in groups 1 and 2 with the six values of the winsorized percent are fitted to a linear regression model. The summary of the obtained values of the regression coefficient, b for the four fitted regression models each for groups 1 and 2 are presented in **Table 4**.

At *step 2* of the proposed alternative approach, an absolute difference between the obtained regression coefficients (i.e., the slope) for each variable in group 1 and 2 was also obtained. The summary of the absolute values is presented as the last row of **Table 4**.

The two step approach of the proposed alternative method was repeated using the data of **Table 3**. The summary of the obtained values of the regression coefficient, b for the three fitted regression models each for groups 1 and 2, and the summary of the absolute values is presented as the last row of **Table 5**.

A cursory look at **Table 4** shows that the regression coefficient (−2.316) of X_2 in group 1 is notably lower than the corresponding regression coefficient (−0.569) of variable X_2 in group 2, and the regression coefficients obtained for the variables X_1, X_3, and X_7 in group 1 and in group 2, respectively. Also, the value of 1.8 which is the absolute difference between the regression coefficients of X_2 in group 1 and in group 2 is significantly greater than the decision boundary value of 0.75. This indicates that the variable X_2 does not have similar variances in groups formed by the dependent, and thus becomes the variable identified with legitimate contaminants. In **Table 5**, the values of the regression coefficients for the three variables in groups 1 and 2 are equivalently equal. Also, the values of the absolute difference between the regression coefficients of variable X_5, X_{10}, and X_{11} in group 1 and in group 2 are all equal, and significantly lower than the

Group	Regression coefficients, *b*		
	X_5	X_{10}	X_{11}
1	0.009	0.009	0.013
2	−0.023	0.024	−0.020
$\delta_{abs} = b_{1i} - b_{2i}$	0.0	0.0	0.0

Table 5.
Regression coefficients, b *of fitted regression model to* ***Table 3*** *data.*

decision boundary value of 0.75. This indicates that there are no legitimate contaminants in variable X_5, X_{10}, and X_{11}, respectively. This therefore implies that the fit between the validation sample, D_N^t and the basic assumptions of PDA is sufficient to construct a PDF whose hit rate can be said to be statistically optimal.

6. Conclusions

This paper addresses the issue of achieving a statistically optimal classification accuracy in PDA by first achieving an optimal training sample. For the first real dataset, a training sample of four variables was obtained using the SPSS stepwise method. The training sample gave a hit rate of 86.0%. When all legitimate contaminants in one of the four variables have been identified and eliminated using the MW-GD method, an optimal training sample was achieved. The optimized training sample was used to construct the PDF, Z_{opt} (13) which gave a classification accuracy of 97.0% when tested on the initial training sample of four variables. This significant increase in classification accuracy suggests that the use of the WM-GD method seems to effectively enhance the similarity of each predictor variable variances between groups, thus taken into account the basic assumptions needed to achieve a statistically optimal classification accuracy. Using the second real dataset, a training sample of three variables was obtained and used to construct the PDF, Z (16), which yielded 85.0% hit rate. When the modified mean values of the three variables were plotted, the bar shape of the three variables in the two groups was similar. This means that the PDF, Z (16) hit rate of 85.0% cannot be increased further because the training sample that gave birth to it was statistically optimal.

The uniqueness of the MW-GD method lies in its ability to effectively identify and eliminate legitimate contaminants in one or several predictor variables, thus resolving any significant differences in the variances of the predictor variables between the groups. In other words, the MW-GD method is unique in its ability to sufficiently account for the basic assumptions required to achieve statistically optimal classification accuracy in PDA. As a result, an optimal training sample obtained from the first real dataset gave a statistically optimal hit rate of 97.0% compared to an initial maximum hit rate of 86.0%. For the second real dataset, the method was successful in confirming the optimality of the initial training sample obtained using the SPSS stepwise method. Similarly, the graphical diagnostic was able to identify the predictor variable(s) whose variance was not similar within the groups. Consequently, the graphical diagnostic associated with the proposed method could be used as an alternative graphical test of homogeneity of variances in PDA.

Another important contribution to the MW-GD method in this paper was the proposed alternative statistical interpretation for the graphical diagnostic associated with the MW-GD method demonstrated in Subsection 5.2. This proposed alternative statistical interpretation proved very effective in terms of identifying the variable with legitimate contaminants, and could serve as a useful alternative tool for identifying variables with legitimate contaminants in the event of any difficulties in differentiating between a variable shape in the groups of the 2-D area plot.

Finally, two real training samples have been used. Consequently, the validity of the experimental results is limited to the scope of the datasets used. Therefore, this paper believes that more experimental results are needed in order to reach a final conclusion on the efficiency of the MW-GD method compared to classical alternatives known to improve classification accuracy in PDA.

Nomenclature

D_N^t	this is the complete list of objects used in building the predictive discriminant function (PDF)
D_N^t	this is the complete list of objects without outliers and hidden influential observation used in building the predictive discriminant function (PDF)
K	the number of groups or categorical criterion
N	total number of cases over all groups in a dataset
N^t	total number of cases over all groups in the optimized training sample
n_v	the number of validation samples
P	the number of predictor variables
$P^{(a)}$	this is the percentage of cases on the diagonal of the confusion matrix, or simply the percent of correct classification
$\hat{p}^{(a)}$	this is the hit-rate obtained by applying a rule based on a particular training sample to future samples taken from the same population
$\hat{p}^{(a)}_{LOOCV}$	this is the leave-one-out cross-validation (LOOCV) estimate of the optimized hit rate
X_i	predictor variables
u_i	discriminant weights
Z	a predictive discriminant function (PDF) created by a linear combination of observable variables
$Z_{(Optimal)}$	an optimized predictive discriminant function (PDF) created by a linear combination of observable variables without outliers and hidden influential observations
Z_j	value for the jth observation in the optimized training sample
$\hat{Z}_j$	predicted response for the jth observation in the optimized training sample
$\hat{Z}^{-j}$	predicted response for the jth observation calculated with the jth observation removed from the training sample

Author details

Augustine Iduseri
Department of Statistics, University of Benin, Benin City, Nigeria

*Address all correspondence to: augustine.iduseri@uniben.edu

References

[1] Huberty CJ, Olejnik S. Applied Manova and Discriminant Analysis. Hoboken, New Jersey: John Wiley and Sons Inc.; 2006. p. 406

[2] Iduseri A, Osemwenkhae JE. On estimation of actual hit rate in the categorical criterion predicting process. Journal of the Nigerian Association of Mathematical Physics. 2014;**28**(1):461-468

[3] Huberty CJ. Applied Discriminant Analysis. New York: Willey and Sons; 1994

[4] Thompson B. Stepwise regression and stepwise discriminant analysis need not apply here: A guidelines editorial. Educational and Psychological Measurement. 1995;**55**(4):525-534

[5] Uray M. Incremental, robust, and efficient linear discriminant analysis learning [thesis]. Graz, Austria: Institute for Computer Graphics and Vision, Graz University of Technology; 2008

[6] Iduseri A, Osemwenkhae JE. A new approach for improving classification accuracy in predictive discriminant analysis. Annals of Data Science. 2018; **5**(3):339-357. DOI: 10.1007/s40745-018-0140-9

[7] Croux C, Filzmoser P, Joossen K. Classification efficiencies for robust linear discriminant analysis. Statistica Sinica. 2008;**18**:581-599

[8] Draper NR, Smith H. Applied Regression Analysis. New York: Wiley; 1981

[9] Huberty CJ. Problems with stepwise methods: Better alternatives. In: Thompson B, editor. Advances in Social Science Methodology. Vol. 1. Greenwich, CT: JIA Press; 1989. pp. 43-70

[10] Bertrand C, Ernest F, Hao HZ. Principles and Theory for Data Mining and Machine Learning. Springer Series in Statistics. New York: Springer; 2009. pp. 569-576. DOI: 10.1007/978-0-387-98135-2

[11] Chiang LH, Russell EL, Braatz RD. Fault Detection and Diagnosis in Industrial Systems. New York: Springer; 2001

[12] Hand DJ. Branch and bound in statistical data analysis. Journal of the Royal Statistical Society: Series D (The Statistician). 1981;**30**(1):1-13

[13] Siedlecki W, Sklansky J. A note on genetic algorithms for large-scale feature selection. Pattern Recognition Letters. 1989;**10**(50):335-347

[14] Huerta EB, Duval B, Hao J. A hybrid LDA and genetic algorithm for gene selection and classification of microarray data. Neurocomputing. 2010;**73**: 2375-2383

[15] Tibshirani R. Regression shrinkage and selection via the LASSO. Journal of the Royal Statistical Society: Series B: Methodological. 1996;**58**(1):267-288. DOI: 10.1111/j.2517-6161.1996.tb02080.x

[16] Shih-Wei L, Shih-Chieh C. A particle swarm optimization approach for enhancing classification accuracy rate of linear discriminant analysis. Applied Soft Computing. 2009;**9**(3):1008-1015. DOI: 10.1016/j.asoc.2009.01.001

[17] Jiang Z, Lin Z, Davis LS. Learning a discriminative dictionary for sparse coding via label consistent K-SVD. In: Proceedings of the 24th IEEE International Conference on Computer Vision and Pattern Recognition (CVPR); Colorado Springs, CO, USA; 20–25

June 2011. pp. 1697–1704. DOI:10.1109/CVPR.2011.5995354

[18] Yang M, Zhang L, Feng X, Zhang D. Fisher discrimination dictionary learning for sparse representation. In: Proceedings of the IEEE International Conference on Computer Vision (ICCV); Barcelona, Spain; 6–13 November 2011. pp. 543–550. DOI: 10.1109/ICCV.2011.6126286

[19] Kong S, Wang D. A brief summary of dictionary learning based approach for classification [Internet]. 2012. Available from: http://arxiv.org/pdf/1205.6544 [Accessed: 2021-12-06]

[20] Mary-Huard T, Robin S, Daudin J. A penalized criterion for variable selection in classification. Journal of Multivariate Analysis. 2007;**98**:695-705. DOI: 10.1016/j.jmva.2006.06.003

[21] Daud M, Muhammad A, Affindi B, Retno V. Improving classification algorithm on education dataset using hyperparameter tuning. Procedia Computer Science. 2022;**197**:538-544. DOI: 10.1016/j.procs.2021.12.171

[22] Naman SB, Abhishek DP, Jegadeeshwaran R, Kaushal AK, Rohan SG, Atharva MK. A Bayesian optimized discriminant analysis model for condition monitoring of face milling cutter using vibration datasets. Journal of Nondestructive Evaluation. 2022;**5**(2): 021002. DOI: 10.1115/1.4051696

[23] Osemwenkhae JE, Iduseri A. Efficient data-driven rule for obtaining an optimal predictive function of a discriminant analysis. Journal of the Nigerian Association of Mathematical Physics. 2011;**18**:373-380

[24] Stąpor K, Smolarczyk T, Fabian P. Heteroscedastic discriminant analysis combined with feature selection for credit scoring. Statistics in Transition New Series. 2016;**17**(2):265-280

[25] Iduseri A, Osemwenkhae JE. An efficient variable selection method for predictive discriminant analysis. Annals of Data Science. 2015;**2**(4): 489-504

[26] Nguyen HV, Gopalkrishnan V. Feature extraction for outlier detection in high-dimensional spaces. Journal of Machine Learning Research. 2010;**10**(2): 252-262

[27] Alrawashdeh MJ, Muhammad Sabri SR, Ismail MT. Robust linear discriminant analysis with financial ratios in special interval. Applied Mathematical Sciences. 2012;**6**: 6021-6034

[28] Lim YF, Syed Yahaya SS, Idris F, Ali H, Omar Z. Robust linear discriminant models to solve financial crisis in banking sectors. In: Proceedings of the 3rd International Conference on Quantitative Sciences and Its Applications; Langkawi, Kedah; 12–14 August 2014. pp. 794–798. DOI:10.1063/1.4903673

[29] Syed Yahaya SS, Lim Y, Ali H, Omar Z. Robust linear discriminant analysis. Journal of Mathematics and Statistics. 2016;**12**(14):312-316. DOI: 10.3844/jmssp.2016.312.316

[30] Osborne J, Amy O. The power of outliers (and why researchers should always check for them). Practical Assessment, Research and Evaluation. 2004;**9**(6):1-8

[31] Campbell NA. Shrunken estimators in discriminant and canonical variate analysis. Journal of the Royal Statistical Society: Series C: Applied Statistics. 1980;**29**(1):5-14. DOI: 10.2307/2346404

[32] Campbell NA. Robust procedures in multivariate analysis II: Robust canonical variate analysis. Journal of the Royal Statistical Society: Series C: Applied Statistics. 1982;**31**(1):1-8. DOI: 10.2307/2347068

[33] Gomez MJ, DeBenzo Z, Gomez C, Marcano E, Torres RH. Comparison of methods for outlier detection and their effects on the classification results for a particular data base. Analytica Chimica Acta. 1990;**239**:229-243

[34] Critchley F, Vitiello F. The influence of observations on misclassification probability estimates in linear discriminant analysis. Biometrika. 1991; **78**:677-690

[35] Sadek RF. Influence of outliers in classification analysis [thesis]. Anthens: University of Georgia; 1992

[36] Fung WK. On the equivalence of two diagnostic measures in discriminant analysis. Communications in Statistics - Theory and Methods. 1998;**27**:1923-1935. DOI: 10.1080/03610929808832199

[37] Riani M, Atkinson AC. A unified approach to outliers, influence, and transformations in discriminant analysis. Journal of Computational and Graphical Statistics. 2001;**10**(3): 513-544. DOI: 10.1198/106186001317114965

[38] Acuña E, Rodríguez C. An empirical study of the effect of outliers on the misclassification error rate. IEEE Transactions on Knowledge and Data Engineering. 2004;**17**:1-21

[39] Mahalanobis PC. On the generalized distance in statistics. In: Proceedings of the 12th National Institute of Science; India. 1963. pp. 49-55

[40] Iglewicz B, Hoaglin DC. How to Detect and Handle Outliers. ASQC Basic References in Quality Control. Milwaukee, Wis: ASQC Quality Press; 1993

[41] Myatt GJ. Making Sense of Data: A Practical Guide to Exploratory Data Analysis and Mining. U. S. A.: A John Willey and Sons, Inc., Publication; 2007

[42] Lachenbruch PA. Discriminant Analysis. New York: Hafner; 1975

[43] Klecka WR. Discriminant Analysis. London: Sage Publications, Berverly Hills; 1980. p. 61

[44] Ashcraft AS. Ways to evaluate the assumption of multivariate normality. In: Paper Presented at the Annual Meetings of the Southwestern Psychological Association; New Orleans, LA. 1998

[45] Tabachnick BG, Fidell LS. Using Multivariate Statistics. 5th ed. USA: Pearson Education, Inc.; 2007. p. 382

[46] Hayes AF, Cai L. Using heteroskedasticity-consistent standard error estimators in OLS regression: An introduction and software implementation. Behavior Research Methods. 2007;**39**(40):709-722

[47] Field A. An Adventure in Statistics: The Reality Enigma. London: SAGE Publications Ltd.; 2016. pp. 315-321

Chapter 12

Mode Interpretation of Aerodynamic Characteristics of Tall Buildings Subject to Twisted Winds

Lei Zhou and Kam Tim Tse

Abstract

Hills alter wind properties not only by accelerating the wind flow but also by changing the flow direction. Therefore, the wind profile near mountainous terrain continuously exhibits both wind speed and wind direction variation along with the vertical height, and is generally referred to as a twisted wind profile (TWP). The aerodynamic properties of the tall building exposed to TWP are significantly different from those exposed to a conventional wind profile (CWP) and thus generate the twisted-wind effect. The spatial–temporal aerodynamic feature of a tall building under unsteady wind is highly complicated, and the associated flow field is a random and obscure high-dimensional dynamic system. To elucidate the fundamental mechanism involved in TWP, principal orthogonal analysis (POD) is employed to identify the pressure and flow patterns. Moreover, the extracted modal features can be used to physically interpret the coherent structure and dynamical patterns hidden in the surface pressure field or the turbulent flow field. This mode interpretation of aerodynamic characteristics of tall building provides a better understanding of the underlying mechanism of the twisted-wind effect.

Keywords: mode interpretation, aerodynamic properties, pressure pattern, flow pattern, spatial–temporal feature, twisted-wind effect, POD; tall building

1. Introduction

Due to nonuniform geometric features of mountainous topography (e.g., slope and curvature), hills can significantly modify wind properties not only by accelerating the wind flow but also by altering the flow direction. As a result, wind profiles in the proximity of hills exhibit both wind speed and direction variation with height, which are generally referred to as twisted wind profiles (TWP), as shown in **Figure 1** [1–4]. Tall buildings built in hilly landforms, such as in Hong Kong and Japan, have a high probability of being attacked by the topography-induced twisted wind. The varying wind directions in a twisted wind profile can produce a highly non-uniform flow field around a building. Exhibiting varying degrees of flow separation and reattachment as well as

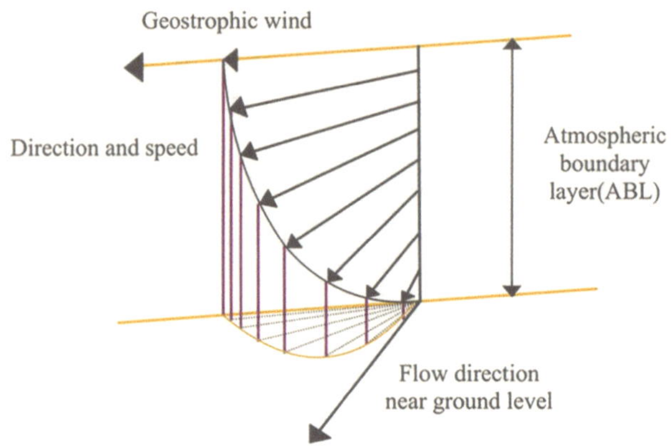

Figure 1.
Schematics of twisted wind (TWP).

varying vortex structures along with the building height. A non-uniform flow field can induce irregular pressure distribution on the external walls of a building and, thus, asymmetric wind loads. Asymmetric wind loads can increase the torsional loading and hence affect sway and twist responses, which are usually only minor considerations when designing buildings in conventional wind profiles. In addition, an increased correlation between wind load components will occur because the wind turbulence associated with a twisted wind profile spans from the windward face to the side face along with the height of a building as a result of the changing wind angles [5–8]. Thus, in comparison to the conventional wind profile (CWP), the aerodynamic characteristics and flow field in the presence of TWP become more complicated, forming a random and nonlinear high-dimensional dynamic system with obscure and elusive features [9, 10].

Therefore, it is of great importance to conduct a deep and systematic investigation on the pressure patterns and flow patterns of tall buildings exposed to the twisted wind instead of only focusing on the micro aerodynamic features but neglecting to reveal the mechanism. This approach is expected to provide a better understanding and physical interpretation of the twisted-wind effect; thus, it can further provide theoretical guidance to ensure wind safety and to optimize the control measures of wind-induced responses for tall buildings built in mountainous terrain.

To extract the spatial-spectral features and dynamic characteristics from the random pressure field, reduced-order models (ROMs) are recommended as an effective way [11, 12]. The goal of ROMs is to search for a relatively straightforward low-dimensional system to represent a chaotic, high-dimensional system through a process of decomposition, truncation, and error estimation [13]. As one of the most typical and effective ROMs models, Proper Orthogonal Decomposition (POD) is a multivariate statistical technique that aims to extract the dominant patterns from the intercorrelated and dependent original observations [14]. The principle of POD is to find a set of the optimal orthogonal bases called principal components in a second-order statistic sense, and then describe the important information by the superposition of the product of the POD base and modal coefficients [15]. POD has numerous advantages like extracting the most important information, compressing the data dimensionality, eliminating noise interference, and more importantly, identifying the structures and patterns contained in the seemingly disordered original observations. As a result, POD has been widely and successfully applied in the fields of both fluid mechanics and wind engineering to capture coherent structures and pressure patterns contained in the flow field and the associated pressure field [16–19]. However, almost all previous studies have only addressed the topic regarding modal identification of the aerodynamic characteristics of tall buildings exposed to conventional wind profiles, Barely no study to date examines that under TWP [20, 21]. It is evident that the aerodynamic properties under TWP are significantly changed and they are not be simply equivalent to the case of CWP with a certain wind incident angle. Thus, it is necessary to identify the coherent structure and interpretate the structures hidden in the aerodynamic features of a tall building, specifically for twisted wind. Additionally, the POD technique is believed to facilitate deep understanding and physical insight into the unique flow and pressure patterns in the presence of twisted wind. This mode interpretation tool is thus helpful to elucidate the underlying aerodynamic interaction mechanism between TWP and tall buildings.

The remaining parts of this paper are organized as follows. Section 2 introduces the methodology of POD and the related notion/notations. Section 3 describes how we used wind tunnel testing and numerical simulation methods to obtain wind pressure and flow field data; Section 4 illustrates the POD analysis results on the wind pressure, and compares the pressure pattern between CWP and TWP cases; Section 5 shows the POD analysis results on the flow field, and compares the flow patterns between CWP and TWP cases. Section 6 gives concluding remarks on the mode interpretation of the aerodynamic characteristics of tall buildings subject to twisted winds as well as recommendations on future work.

2. Methodology

A brief description of the POD principle for the velocity/pressure field is presented here, and more details about the theoretical derivation and practical application can be found in [19].

For moment t_i, velocity or pressure components $f_i(x_m)$ at a monitored location m are arranged in a vector q_i to formulate a snapshot, the total number of velocity snapshots N with sampling interval Δt can then be assembled in matrix Q_N. The Time-average field $\overline{Q}$ should be firstly subtracted from the instantaneous velocity or pressure field to obtain the zero-centered fluctuating matrix Q_N'.

$$q_i = \begin{bmatrix} f_i(x_1) \\ f_i(x_2) \\ \vdots \\ f_i(x_m) \\ \vdots \\ f_i(x_M) \end{bmatrix} \quad (1)$$

$$Q_N = [q_1, q_2, q_3 \cdots q_N] \quad (2)$$

$$Q_N' = Q_N - \overline{Q} \quad (3)$$

Where $\{x_1 \ x_2 \ \cdots \ x_m\}$ signify the position of the monitored mesh points, and M is the total number of the monitored points.

The purpose of POD is to determine a set of orthogonal bases, and equivalently express q_i' as the superposition of the product of the POD base and corresponding modal coefficients.

$$q_i' = \sum_{k=1}^{n} a_{ki}\varphi_k \tag{4}$$

To identify POD basis, a covariance matrix C is formulated as $C = {Q_N'}^T Q_N'$, and the eigenvalue and eigenvector can be obtained by solving the eigenfunction of matrix C as follows,

$$C\varphi_k = \lambda_k \varphi_k \tag{5}$$

where the eigenvalue λ_k reflects the energy contribution of each POD mode and is arrayed in descending sequence, eigenvector φ_k represents the POD modes and any two of them are orthogonal to each other spatially.

Modal coefficients $A = [a_{ki}]\ (k = 1, 2, 3\cdots n;\quad i = 1, 2, 3\cdots N)$ are determined by projecting the original fluctuating velocity or pressure field onto the POD modes as follows,

$$A = \Phi^T Q_N' \tag{6}$$

where the spatial POD mode $\Phi = [\varphi_1 \quad \varphi_2 \quad \dots \quad \varphi_n]$ and the temporal modal coefficient $A = [a_{ki}]\ (k = 1, 2, 3\cdots n;\quad i = 1, 2, 3\cdots N)$.

3. Data resources

3.1 Wind pressure data resources

Pressure measurement testing was carried out on a rectangular tall building, of cross-Section 180 mm × 60 mm (Breadth×Depth) and height of H = 600 mm. The pressure distribution on the 1:400 scaled building model was measured using a DSM3400 synchronous multi-pressure sensing system (SMPSS). **Figure 1** shows arrangement details of the pressure taps and experimental setup. Pressure taps were arranged in 14 rows ×12 columns on the windward/leeward surfaces and in 14 rows ×4 columns on side surfaces. The sampling frequency and duration were 330 Hz and 90 s respectively. In this study, the approaching wind perpendicularly attacking the relatively wider and narrower surface of the building correspond to two different working conditions, i.e., angle of attack (AOA) of 0 degree and 90 degrees respectively.

The three targeted wind profiles have similar distributions of velocity and turbulence but different twisted angle profiles. Specifically, the wind speed and turbulence intensity profiles follow power-law functions with exponents of 0.11 and − 0.24. The maximum twisted angles are 0°, 15° and 30°, thus the profiles are respectively labeled CWP, TWP15 and TWP30. It should be noted that variation of twisted angles with height for TWP15 and TWP30 conforms with the negative power curve expressed in Eqs. (7) and (8), and as given in [22].

$$\theta_{15}(z) = 15 \times \exp\left(-0.0976 \times (z/25)\right) \tag{7}$$

$$\theta_{30}(z) = 30 \times \exp\left(-0.106 \times (z/25)\right) \tag{8}$$

where $\theta_i(z)$ represents the twisted angle measured at the height z; the subscript i denotes the maximum twisted angle of TWP, herein, i = 15 and 30.

Three types of wind fields (CWP, TWP15 and TWP30), were generated in a wind tunnel by using passive simulating facilities. As shown in **Figure 2(b)** and **(c)**, to replicate the twisted wind, the wooden vane system was placed 5 m upstream away from the central axis of the building model, and the windward ends of the single wooden vane were isolated 1 m from each other. It should be noted that the wooden vane employed in this study was adjustable, indicating that the wooden strip at a certain height could be offset at any desired angle. Moreover, to compensate for the dissipation near the wind tunnel ground, three rows of roughness elements were rotated at a certain angle are set in a stagger arrangement and placed between the wooden vane

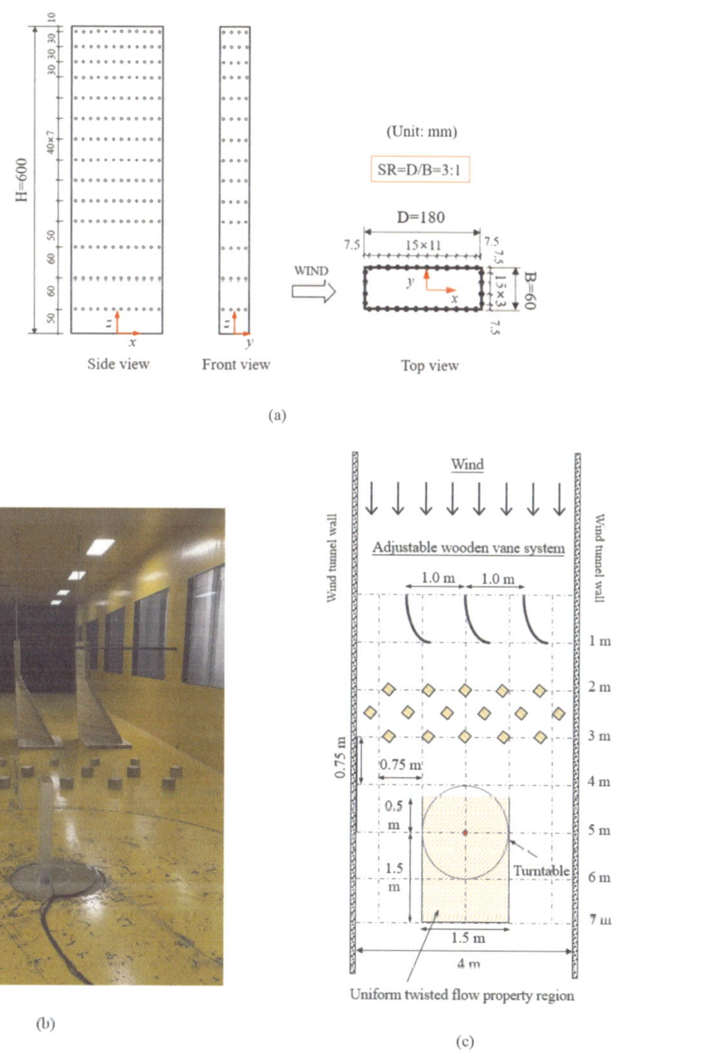

Figure 2.
Experimental setup and details: (a) distribution of pressure taps on building model; (b) TWP test arrangement in wind tunnel; (c) section plan of the experimental setup.

system and the building model. To determine the region with uniform twisted flow properties, a grid measurement system constituted by blue dash lines was utilized to monitor the flow features at each grid point. After trial and error, it turned out that the targeted twisted wind can maintain consistent flow properties within a rectangular region (shaded in orange) with dimensions of 1.5 m × 2.0 m (width ×length) around the turntable center. z_{ref} indicates the reference height which is located at the building roof (z = 0.6 m); U_{ref} denotes the reference velocity and equals 6.2 m/s; I_{ref} represents the reference turbulence intensity and equals 6.9%. According to the reference wind speed U_{ref} and building breath B, the Reynolds number is calculated as $Re = 2.55 \times 10^4$.

3.2 Flow field data resources

3.2.1 Inflow turbulence generation

Appropriately replicating the inflow turbulence features is a requisite for obtaining accurate LES simulation results. In this study, the narrowband synthesis random flow generator (NSRFG) technique is utilized to simulate inflow turbulence by generating time history series of the fluctuating velocity. Turbulence integral scales (see Eqs.(9)–(11)) conform to those in wind codes (AIJ, 2004; ESDU 85020, 2001), which are also identical to the setting given in [7]; the velocity distribution in the frequency domain corresponds to the von Karman spectrum (see Eqs. (12)–(14)), which can reflect typical spectral features of the turbulent ABL flow in a wind tunnel. The superimposition of the zero-mean fluctuating component and the mean wind profile formulate the initial inflow boundary condition. For specific implement procedures, one can refer to the work of [23]. Note that the profiles of wind speed, turbulence intensity and twisted angle are consistent with the setting in the wind tunnel.

$$L_u(\mathrm{z}) = 100 \cdot \lambda_L \cdot \left(\frac{z}{30\lambda_L}\right)^{0.5} \tag{9}$$

$$L_v(\mathrm{z}) = 0.5\left(\frac{\sigma_v}{\sigma_u}\right)^3 L_u(\mathrm{z}) \tag{10}$$

$$L_w(\mathrm{z}) = 0.5\left(\frac{\sigma_w}{\sigma_u}\right)^3 L_u(\mathrm{z}) \tag{11}$$

$$S_u(\mathrm{f}) = \frac{4(I_u U)^2 (L_u/U)}{\left[1 + 70.8(fL_u/U)^2\right]^{5/6}} \tag{12}$$

$$S_v(\mathrm{f}) = \frac{4(I_v U)^2 (L_v/U)\left[1 + 188.4(2fL_v/U)^2\right]}{\left[1 + 70.8(2fL_v/U)^2\right]^{11/6}} \tag{13}$$

$$S_w(\mathrm{f}) = \frac{4(I_w U)^2 (L_w/U)\left[1 + 188.4(2fL_w/U)^2\right]}{\left[1 + 70.8(2fL_w/U)^2\right]^{11/6}} \tag{14}$$

where λ_L is the scaled ratio; $L(z)$, $I(z)$ and $S(z)$ are the turbulence integral scale, turbulence intensity and power spectral respectively. The subscripts u, v and w denote the components of the physical quantities in the along-wind, crosswind, and torsional directions respectively; *f* is the frequency.

3.2.2 Numerical simulation details

The computational domain and boundary conditions are illustrated in **Figure 3**. The building model is scaled to 1/400 with dimensions of 180 mm (breadth) × 60 mm (depth) × 600 mm (height) and set as wall condition. The calculation domain covers the ranges of 20H in the streamwise direction, 16H in the crosswind direction and 8H in the vertical direction. In this study, the wind incident angle of 90 degree exactly corresponds to the side ratio case of 3:1, where the flow phenomenon is more complex.

The structured hexahedral grids were utilized to mesh the outer calculation domain, while for the inner region, an O-shaped grid with good orthogonality was employed. Additionally, local mesh refinement was performed near the building surface and wake region to better capture the significant features of the shear layer and recirculation zone. For example, 10 layers were imposed on the building wall to simulate the near-wall velocity gradient. The normal growth rate of the mesh within the boundary layer was 1.02, while for the grids was relatively far away from the external surface boundary, the growth rate was 1.1. To ensure y^+ in the most region was around 1, the distance from the building wall to the center of the first layer was set as $\Delta y/B = 2.0 \times 10^{-4}$. The mesh scheme and distributions around the building are illustrated in **Figure 4**.

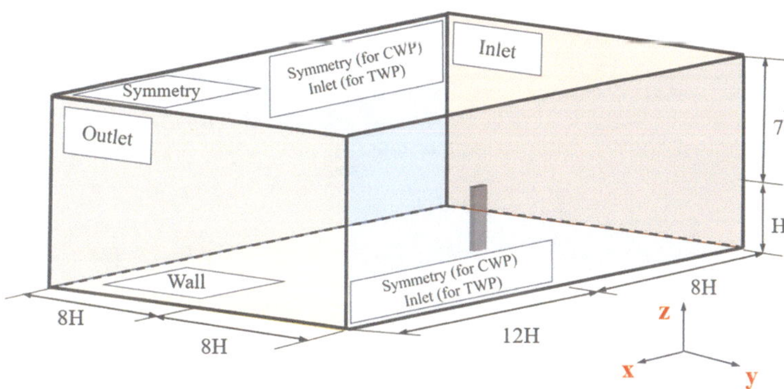

Figure 3.
The boundary conditions of the computational domain for CWP and TWP.

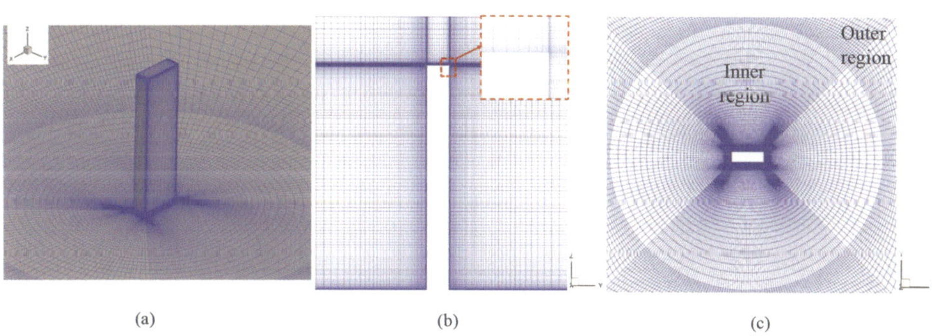

Figure 4.
Mesh scheme and distributions around building: (a) perspective view of 3D mesh distribution; (b) side view of the local 2D mesh (y = 0); (c) top view of the local 2D mesh (z/H = 0.5).

LES with the NSRFG-generated inflow boundary condition was adopted in this study to evaluate the effect of different wind profiles on a tall building. The numerical solver used herein is the commercial CFD code FLUENT. The finite volume method (FVM) was employed to solve the governing equations. SIMPLEC algorithm, initially proposed by [24], was adopted to solve the pressure–velocity coupling scheme. The convection and diffusion terms are discretized using the second-order upwind scheme. The temporal discretization scheme is the implicit time-stepping method. The time step in this paper was set as $\Delta t = 5 \times 10^{-5}$ to ensure the courant number in most calculation region is less than one.

3.3 Validation of the simulated TWP and numerical results

The flow properties of the targeted, experimental and numerical wind profiles at the building location for CWP, TWP15 and TWP30 are compared in **Figure 5**. For better comparison, these three types of wind flow have identical wind speed and

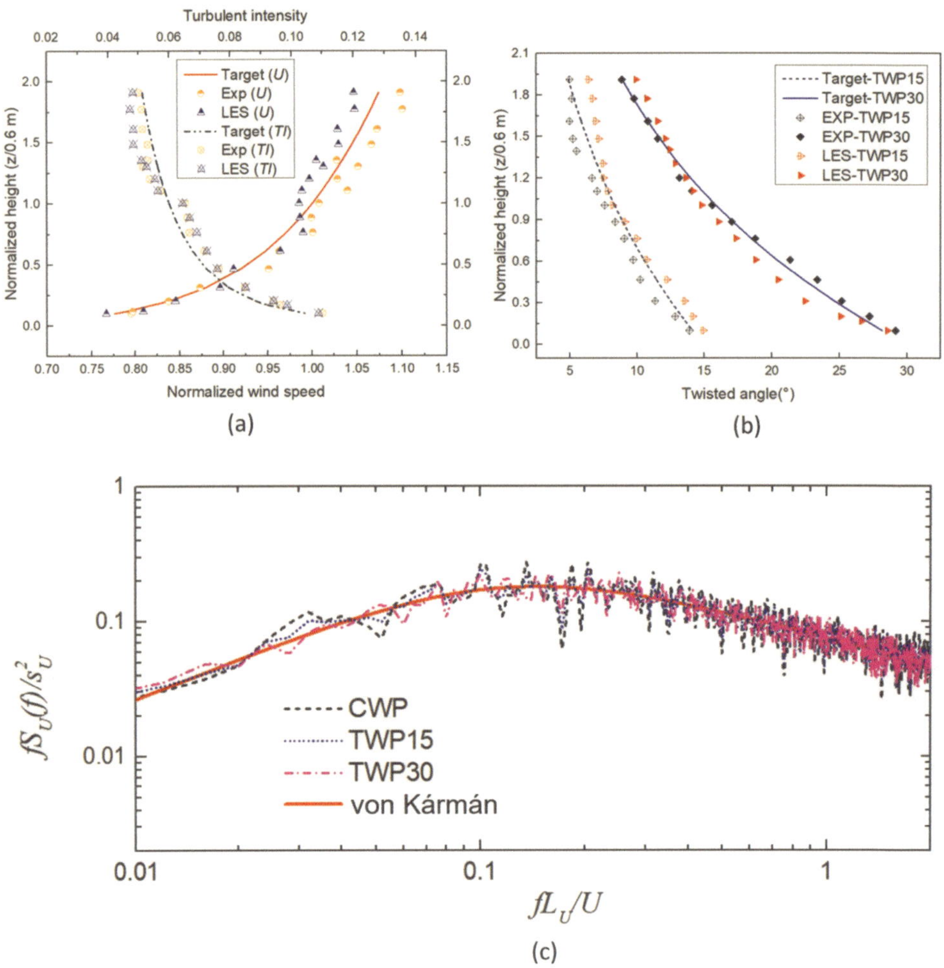

Figure 5.
Flow properties of the targeted, experimental and numerical wind profiles at the location of building for CWP, TWP15 and TWP30: (a) normalized wind speed and turbulent intensity profile; (b) twisted angle profile; (c) power spectrum.

turbulence intensity profile but different twisted angle profiles. Thus, only one set of data is provided in **Figure 5(a)**. Evidently, both the numerically and experimentally replicated wind profiles agree well with the targeted result. The simulated twisted wind conditions, TWP15 and TWP30, had the largest twisted angle at the ground level, which is about 15°and 30°, and both of them well conformed to the targeted variation trend in the vertical direction as expressed in Eqs. (7) and (8). The longitudinal wind velocity spectrum at the reference height is in good agreement with the von Karman spectrum within a relatively wide scope, including the inertial subrange, as shown in **Figure 5(c)**.

The distribution of the pressure coefficients along the circumference of the building at a height of z = 2/3H obtained by LES and wind tunnel test are compared in **Figure 6** for cases CWP and TWP30. Affected by twisted flow, the location of the largest positive pressure and stagnation point is moved towards the right-side of the building on the windward side. When exposed to CWP, the pressure distributions on the left- and right- side surfaces of the building are symmetrical, while under TWP, significant differences are noted both for mean and fluctuating coefficients. The most distinctive discrepancy of the mean pressure between CWP and TWP30 occurs at the right-side surface of the building, simply because it becomes partially windward when the building is exposed to the twisted wind. A similar phenomenon can also be observed for the fluctuating pressure with the largest difference 86.84% appearing at monitoring point 22. Notably, wind twisting can significantly magnify the mean C_p while greatly weakening the fluctuating C_p on the right-side surfaces.

Generally, the pressure distribution obtained in LES at 2/3 height of building shows reasonably good agreement with its counterpart in wind tunnel testing, especially for mean Cp. As a result, although small discrepancies exist, the numerical methods used in this study can still provide reliable and reasonable results for the evaluation of the twisted-wind effect.

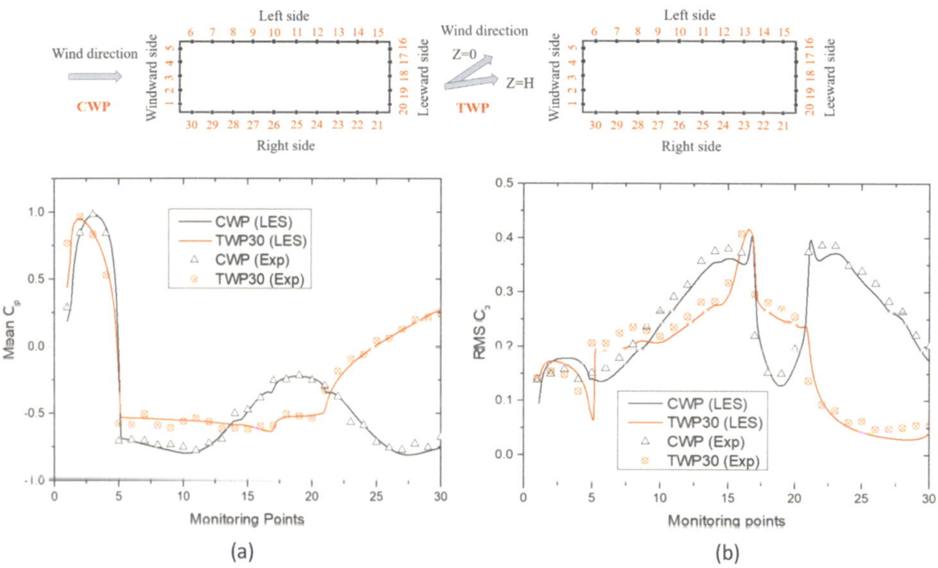

Figure 6.
Mean and fluctuating pressure coefficients at height of z = 2/3H obtained by LES and experiment under CWP and TWP30.

4. POD analysis of surface pressure

The rectangular building with AOA of 0°and 90°represents two different side ratio cases, in which the twisted wind can exert remarkably different effects on the pressure pattern. As a result, to better understand the fluid–structure interaction mechanism, POD was employed to identify the pressure patterns hidden in the fluctuating pressure field on the building surface.

4.1 Mode energy distribution

The energy contributions of the dominant modes and the normalized cumulative number of modes are presented in **Figure 7**. Affected by twisted flow, the fluctuating energy of each pressure mode was reduced to a certain degree, especially for the dominant modes. The energy proportion of the first mode in the case of 0°AOA was larger than that in the case of 90°AOA for the two different flow types CWP and TWP30. Moreover, the difference in the energy percentage between the dominant modes for the case 0° AOA was much more significant in comparison with that of 90° AOA. For example, the fluctuating energy occupied by the first mode of CWP with 0° AOA was around 33%, more than twice that of the second mode. This observation can be attributed to the fact that a tall building with a larger side ratio can produce more complicated flow motion, topology, and behavior. As a result, to reach 90% of the total energy, only the first 5% modes were needed for case 0°AOA, but much larger modes of 24% were needed for the 90° AOA case.

4.2 PSD of the modal coefficients

The power spectral density of the first two POD modes coefficients for the 0° and 90° AOA cases are compared in **Figure 8**. Generally, the twisted flow had a larger effect on the low-frequency region than the high-frequency region for the first mode, while this observation was exactly the opposite for the second mode.

In the case of 0°AOA, as marked by a rectangle in **Figure 8**(a_1), the low-frequency shift mode dominated the first pressure pattern for case CWP while it became less pronounced for case TWP30. The second POD mode of case CWP was distinctively controlled by the vortex shedding phenomenon with the peak appearing at $f_{11} \approx 0.1$.

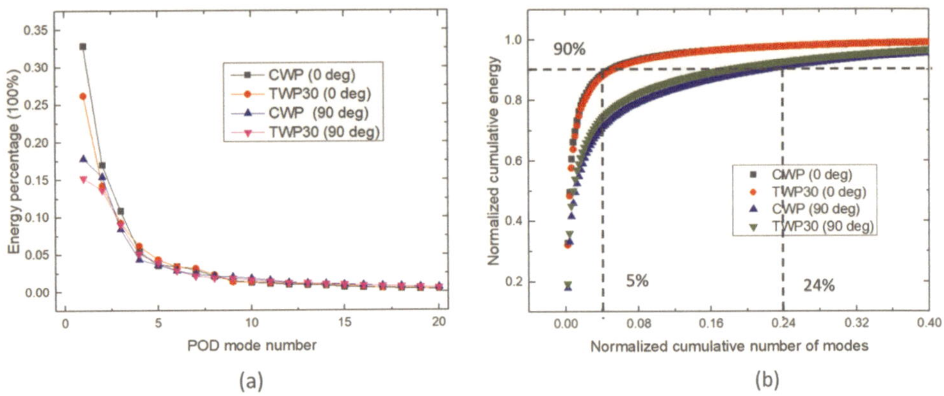

Figure 7.
Energy contributions of (a) dominant modes; (b) normalized cumulative number of modes.

However, the twisted wind reduces this peak energy, indicating that the Karman vortex motion is weakened in the presence of TWP.

In the case of 90°, as shown in **Figure 8**(b_1), although the multi-peak region can be observed in the PSD of the first-order mode for the two wind profiles, it should be noted that the shift mode only appeared in the case of CWP. The second mode under CWP is controlled by two types of vortices shedding behavior with f_{21} = 0.04 and f_{22} = 0.102, and as a result, these two different flow motions around a large aspect ratio building (3:1) can be speculated to be leading-edge vortex shedding (LEVS) and trailing-edge vortex shedding (TEVS) respectively [25]. The existence of twisted flow significantly changes the energy magnitude, the main frequency and the number of peaks. For example, in the case of TWP30, there was only one pronounced peak, whose energy was slightly reduced but the frequency increased to f31 = 0.048. Thus, it is reasonable to surmise that the second mode of TWP 30 is highly related to alternate-edge vortex shedding (AEVS) [25, 26].

4.3 POD pressure pattern

The first two POD modes of the pressure field on the building surface for case CWP and TWP30 with 0°AOA are shown in **Figure 9**. When exposed to CWP, the first mode pattern had a similar distribution with the mean pressure filed on the

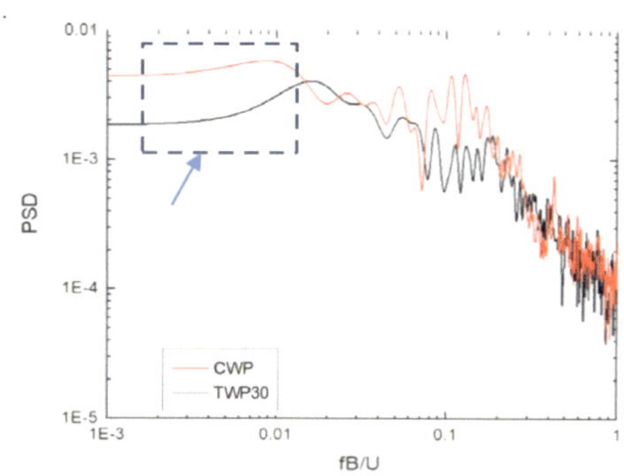

(a_1) The first mode coefficient

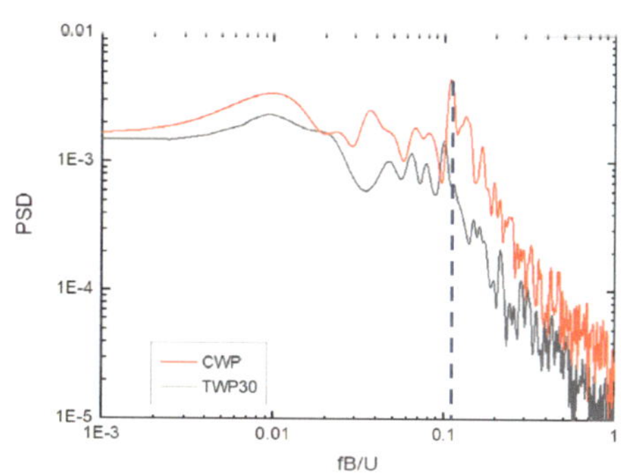

(a_2) The second mode coefficient

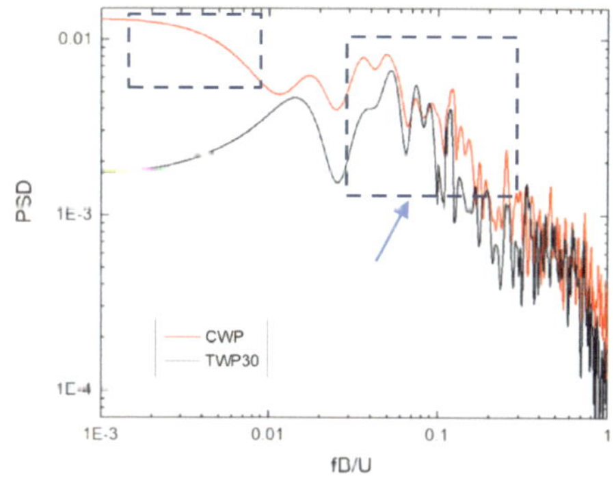

(b_1) The first mode coefficient

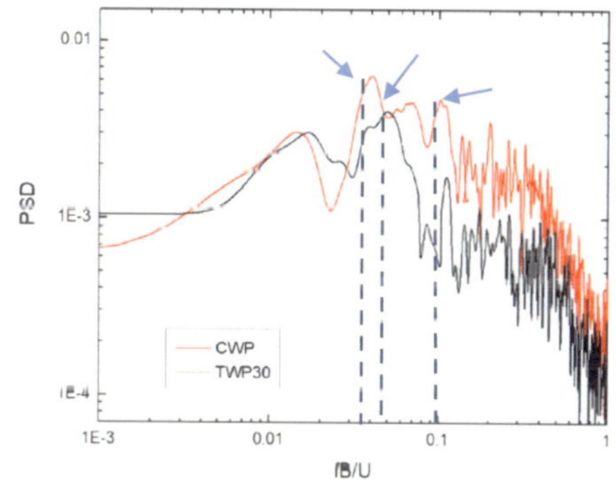

(b_2) The second mode coefficient

Figure 8.
Power spectral density of the first two POD mode coefficients for two AOA cases of (a) 0°; (b) 90°.

building surface as depicted in **Figure 8**, which implies that the first POD mode possibly corresponds to the low-frequency shift mode. Moreover, it is reasonable to deduce that the second asymmetrical mode shape well captured the fluctuating properties induced by the periodic vortex shedding.

In the presence of twisted wind, the POD mode pattern on the building surfaces became asymmetric and non-uniform. It was evident that the vertical varying wind direction had a noticeable influence on the mode shape on the leeward surface. Specifically, the first mode shape was significantly deflected towards the right side near the downstream region (see black dash circle). However, the second mode shape indicated that twisted flow can amplify the positive region near the left-side surface but shrink the negative area near the right-side surface, as indicated by the dashed rectangle in **Figure 9** (a_2) and (b_2). The possible reason could be that the conical vortex near the left side was enhanced but suppressed near the right side on the appearance of TWP.

The first two POD modes of the pressure field on the building surface for case CWP and TWP30 with 90°AOA was shown in **Figure 10**. Under CWP, a shift mode can also be observed for the first POD mode. Notably, the pressure pattern on the left and right sides are symmetrically distributed, but experiences an initially upward and then downward process from the windward edge to the leeward edge. Such finding indicates that flow separation and reattachment simultaneously occur at the side

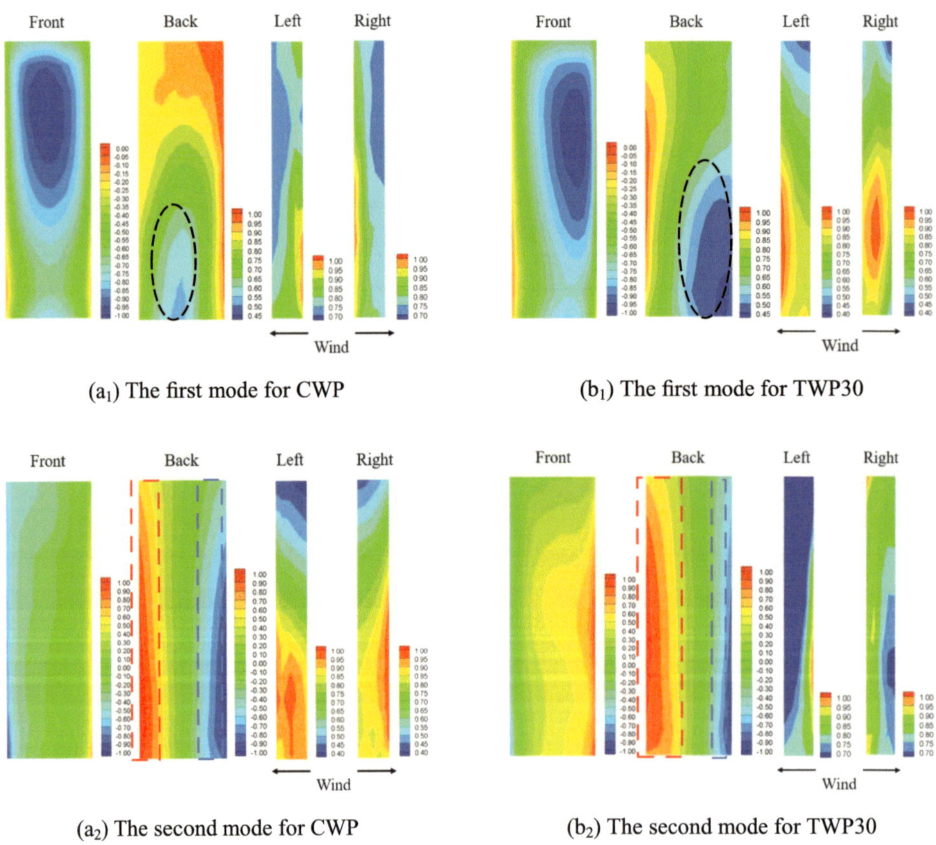

(a_1) The first mode for CWP

(b_1) The first mode for TWP30

(a_2) The second mode for CWP

(b_2) The second mode for TWP30

Figure 9.
The first two POD modes of surface pressure field at 0°AOA for (a) CWP; (b) TWP30.

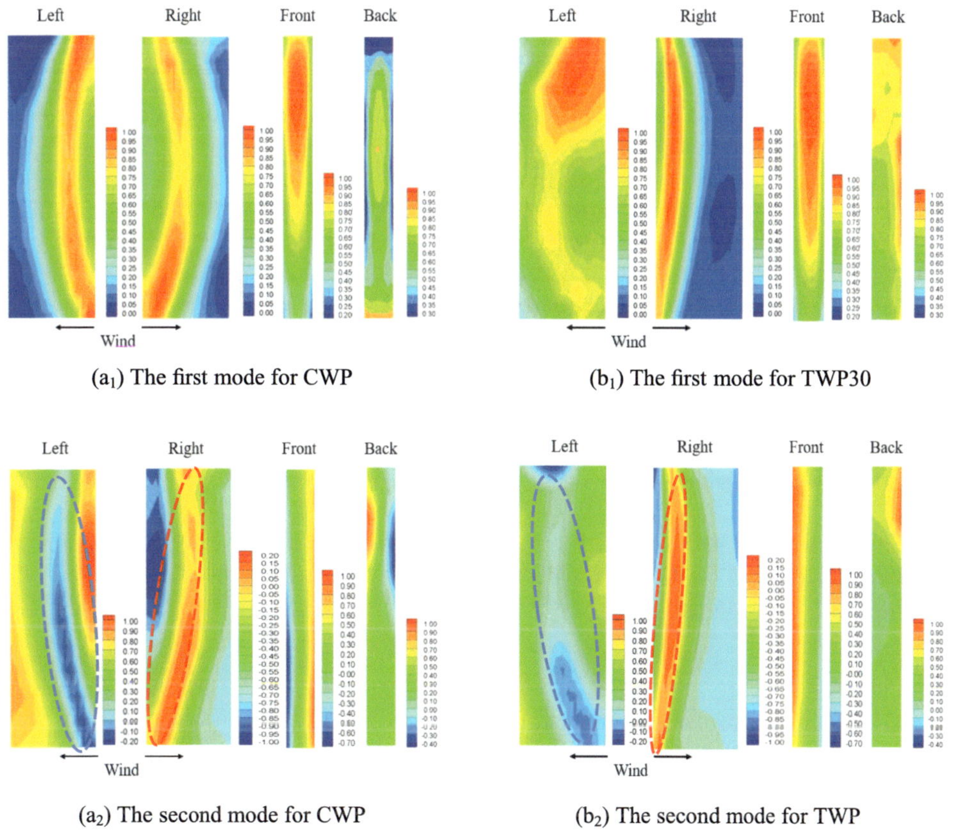

Figure 10.
The first two POD modes of surface pressure field at 90°AOA for (a) CWP; (b) TWP30.

surface of the building with side ratio of 3:1. Similar to the 0°AOA case, the asymmetrical second mode of CWP is strongly related to the vortex shedding phenomenon. Affected by TWP, the continuous varying wind directions along the building height cause the reattachment location on the right-side surface to move upstream (see the red dash circle), however, that on the left-side surface becomes weaker (see blue dash circle). This observation indicates that the conical vortex near the left side is significantly weakened and merges into the wake flow. Overall, TWP has a distinctively different impact on the mode shape of buildings with different aspect ratios.

5. POD analysis of flow field

In this section, the wind incident angle of 0 degree is studied, which exactly corresponds to the side ratio case of 3:1, because in this case, the flow phenomenon is more complex. To identify the mechanisms that underlie the twisted-wind effect and elucidate the essential difference of the flow patterns under CWP and TWP, POD analysis was conducted on fluctuating velocity field at different horizontal planes (z = 1/6H, 1/2H and 5/6H) in the near-wake, and the spatial distribution and temporal evolution of the mode patterns were systematically compared for these two cases.

5.1 Mode energy contribution

Figure 11 shows the energy contributions of the dominant POD modes and cumulative energy of the first three modes. Under CWP, the kinetic energy contribution of the first two POD modes was the most prominent, and both exceeded 15%, while that of the third-order mode sharply decreased to below 10%. The largest cumulative energy proportion of the first three modes (46.71%) appeared at the middle plane z = 1/2H, where the coherent structure was less affected by the free end and the ground effect. This observation is consistent with the finding reported in [20]. Affected by the twisted flow, the energy proportion taken by main POD modes was relatively uniform, especially at the mid-height of the building, and for increasing height, the cumulative energy percentages of the first three main modes gradually decreased from 34.95% to 28.60%.

5.2 POD flow pattern

The first three POD mode patterns of wake velocity field into horizontal plane 1/2 H under CWP and TWP30 are depicted in **Figures 12** and **13** respectively. For CWP,

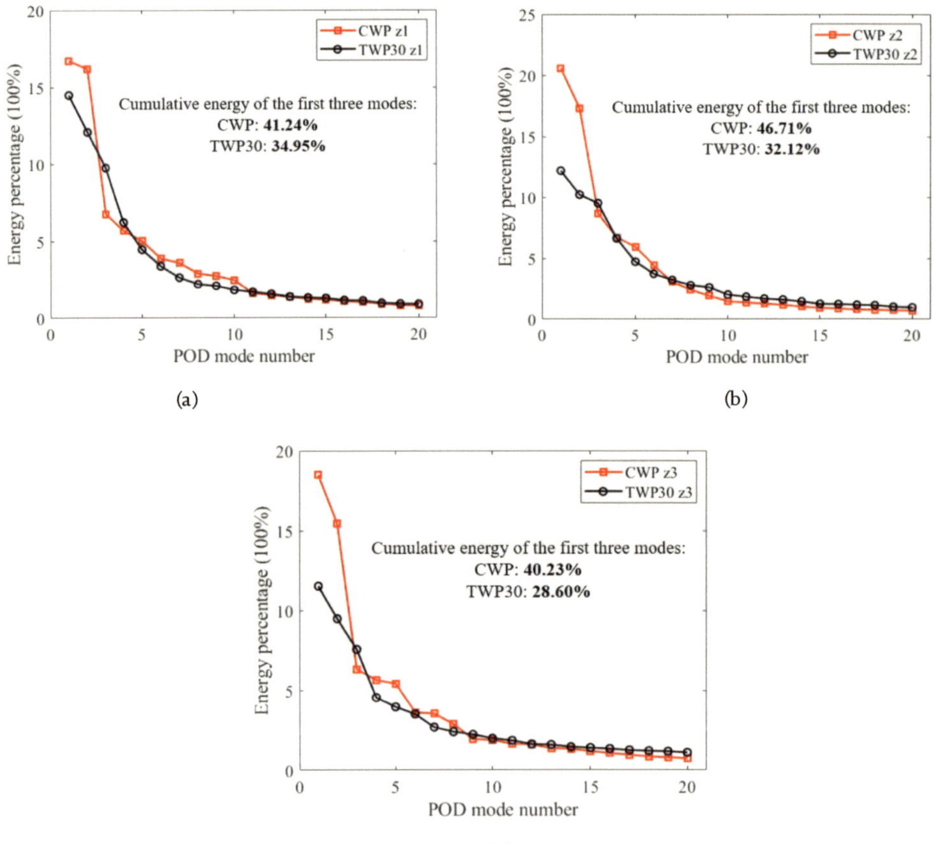

Figure 11.
Energy contributions of dominant POD modes and cumulative energy of the first three modes. (a) z_1=1/6 H. (b) z_2=1/2 H. (c) z_3=5/6 H.

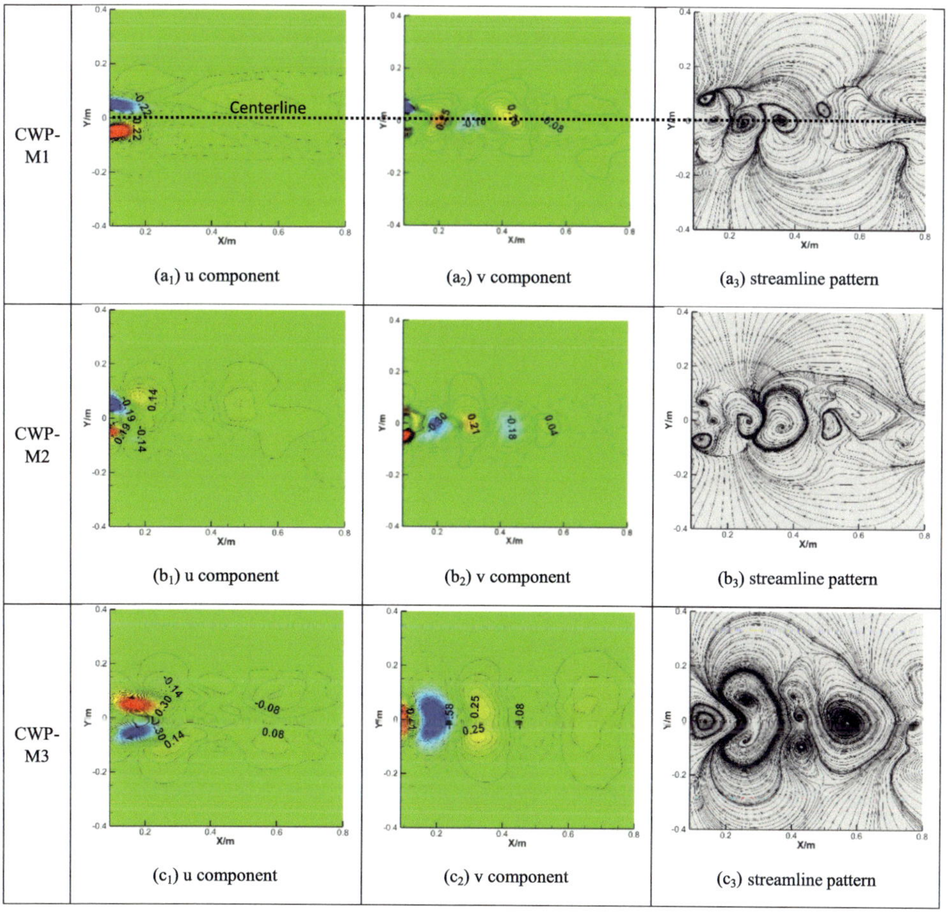

Figure 12.
The first three POD mode patterns of the wake velocity field at horizontal plane of 1/2 H under CWP (a) contour of u component; (b) contour of v component; (c) streamline pattern.

the contour of the u component was asymmetrically distributed while that of the v component was a symmetry to the centerline. The first two POD modes were similar (i.e., streamline pattern) but had an opposite sign, and as a result, can be recognized as a pair of conjugate modes. The third mode was found to have a series of vortices arrayed along the streamwise direction with alternating rotation direction. These observations confirm the findings reported in the previous POD investigation on the wake flow of a cylinder [27, 28].

To further disclose the temporal and spectral characteristics of the main underlying modes in the velocity fields, the PSD of the first three POD mode coefficients at the mid-height plane of the building were analyzed and compared for CWP and TWP30 as shown in **Figure 14**. As can be observed, there were several peaks for the first two POD modes while the most pronounced one appeared at $f11 = 0.04$ and $f12 = 0.11$(see blue arrow), these manifests that although coherent structures were composed of different sizes of vortices, the most dominant types were the LEVS and TEVS. The third mode had the main frequency of 0.11, and more importantly, its modal shape was regularly distributed and closely resembled the Karman vortex. Subsequently, it is reasonable to speculate that the third mode is highly related to the TEVS phenomenon.

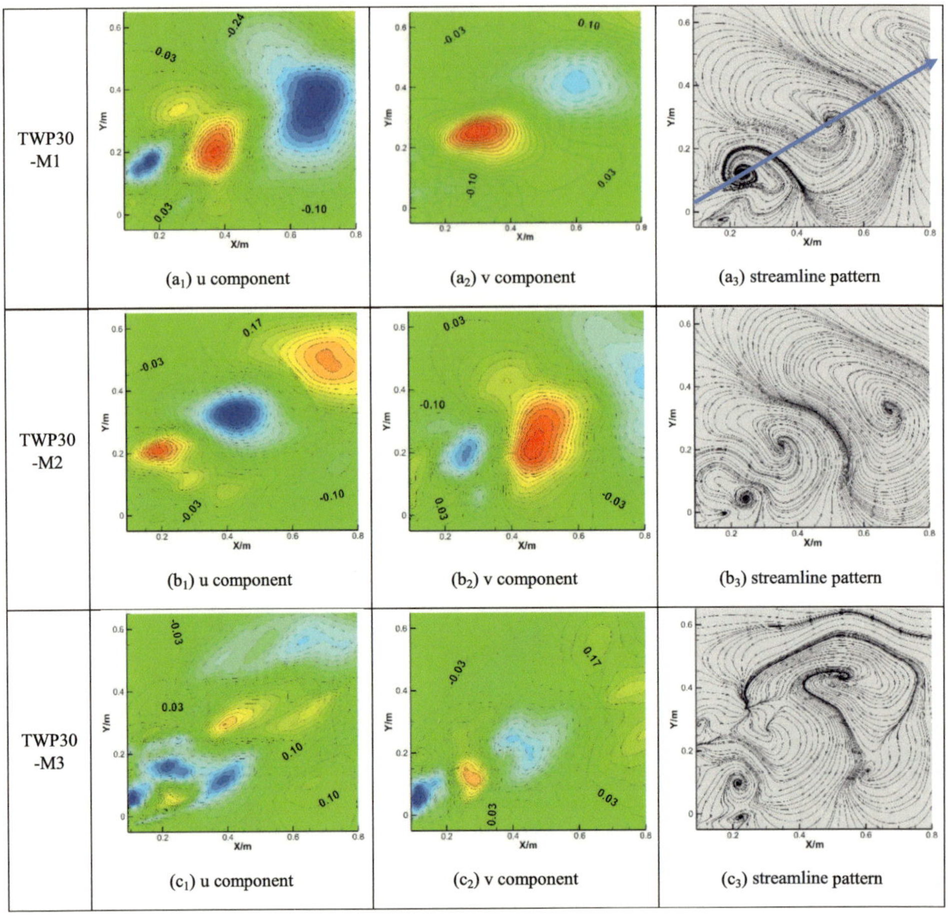

Figure 13.
The first three POD mode patterns of the wake velocity field at horizontal plane of 1/2 H under TWP30 (a) contour of u component; (b) contour of v component; (c) streamline pattern.

Under twisted flow, as shown in **Figure 13**, the mode topology and configuration were both largely modified, moreover, the vortex shedding was no longer aligned with the centerline but deviated towards the wind twisted direction. The vortex structures of the first two modes were very similar along and nearly had the same frequency of f3 = 0.048 (marked by green arrow in **Figure 13(c)**). This finding indicates that a pair of conjugate modes were well captured, with a high possibility to be controlled by the AEVS coherent structure. The twisted flow made the third mode shape less regular than that of CWP, and no distinctive peak was noted in this case.

To fully reveal the POD mode from a three-dimensional perspective, the mode results extracted from the velocity field at another horizontal plane z = 1/6 H under CWP and TWP30 are provided in **Figures 15** and **16**. It was found that the mode results at the plane near to the ground level z = 1/6 H basically shared a similar symmetrical distribution pattern to that of plane z = 1/6 H for the case of CWP and TWP30, despite some discrepancies (e.g., mode shape, magnitude and streamline) still being observed between these two horizontal planes due to the existence of the ground effect. An interesting observation is that because the twisted wind had a maximum twisted angle near the ground level, the twisted degree of the wake

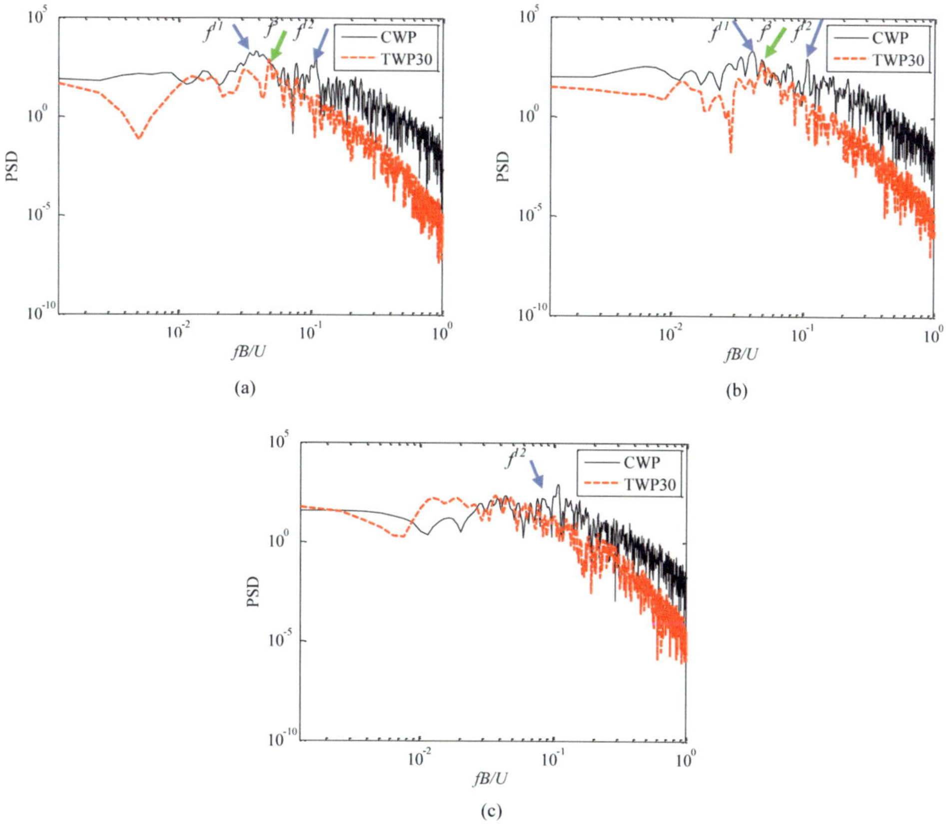

Figure 14.
Power spectral density of the first three modes for CWP and TWP30. (a) Mode 1. (b) Mode 2. (c). Mode 3.

streamline pattern was much larger in the horizontal plane of z = 1/6 H than that of plane of z = 1/2 H, as indicated by the blue arrow in **Figure 13**(a_3) and 16(a_3).

As can be seen in **Figures 15** and **16**, the distribution of mode pattern at the vertical plane x = 0.12 m exhibited symmetrical properties along the centerline y = 0 for the case of CWP, possibly related to the "quadrupole" wake pattern. In the case of TWP30, these POD modes deviated towards the approaching wind twisting direction, namely, the distribution was concentrated in the positive direction of the y-axis. It should be noted that in comparison with the CWP case, the distribution range and amplitude of the POD mode of w velocity were significantly amplified for the case of TWP30. This indicated that the interaction between the vertical velocity components at different heights in the flow field was more intense, which could explain the increase of the aerodynamic correlation of the local wind loads in the vertical direction.

5.3 Correlation of modal coefficients

Figure 17 shows the correlation of the first three POD mode coefficients at two horizontal planes (z_1 = 1/6H and z_2 = 1/2H) for CWP and TWP30.The x- and y-axis represent the modal coefficients of ith mode (i = 1,2 and 3) at the bottom height plane and mid-height plane respectively. For CWP, the phase-plane trajectory of the first three modes all exhibit a circular distribution (see the red dash line), thus, the

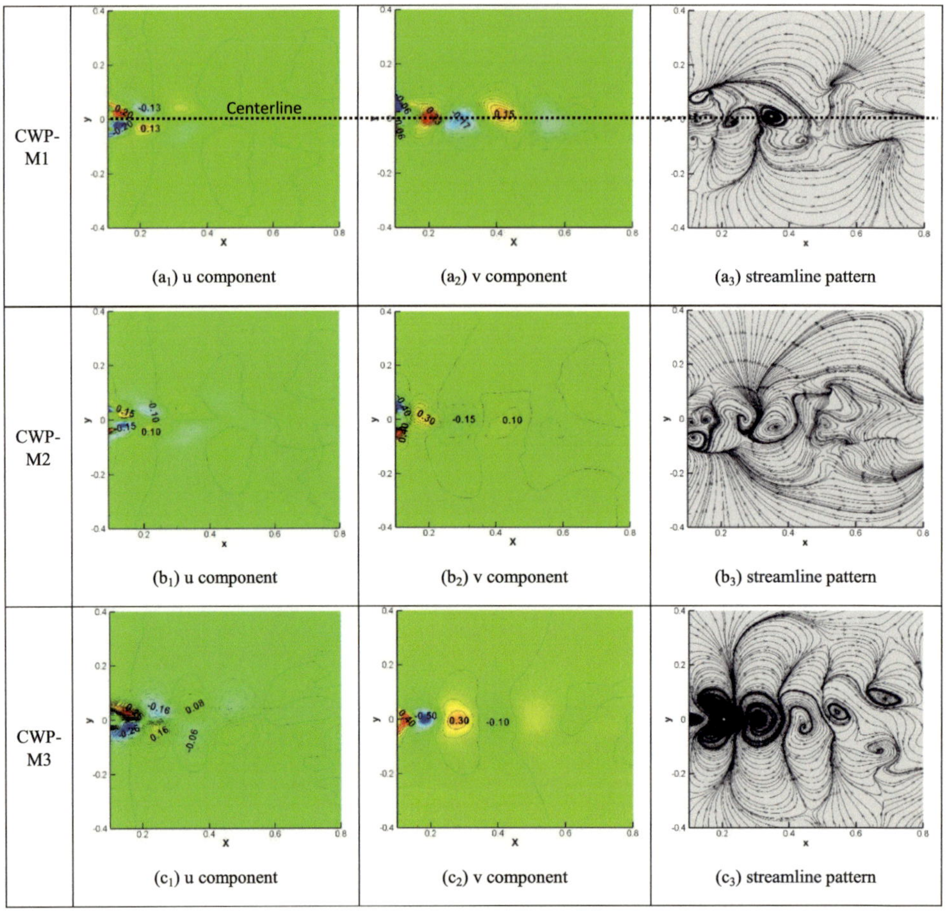

Figure 15.
The first three POD mode patterns of the wake velocity field at horizontal plane of 1/6 H under CWP (a) contour of u component; (b) contour of v component; (c) streamline pattern.

correlation of the same mode at different heights of the building is very close to zero, indicating that the coherent structure at these two locations is relatively independent. Conversely, affected by twisted flow, the distribution of the modal coefficients exhibits an inclined elliptic shape with a largely amplified correlation coefficient (see the black dash line). For example, the first predominant mode at these two different horizontal planes has the largest correlation coefficient *Cor* = 0.50, indicating that the predominant coherent structure simultaneously controls the flow motion at the bottom and middle planes. Notably, the correlation coefficients of the first mode (*Cor* = 0.50) and the second mode (*Cor* = 0.45) are very close to each other, demonstrating the existence of conjunction modes at the middle height of the building z_2 = 1/2H, as demonstrated in **Figure 17**(a_1) and (b_1).

6. Conclusions

This study aims to investigate the essential mechanism of nonlinear dynamic systems, i.e., the random pressure field and the flow field over a tall building

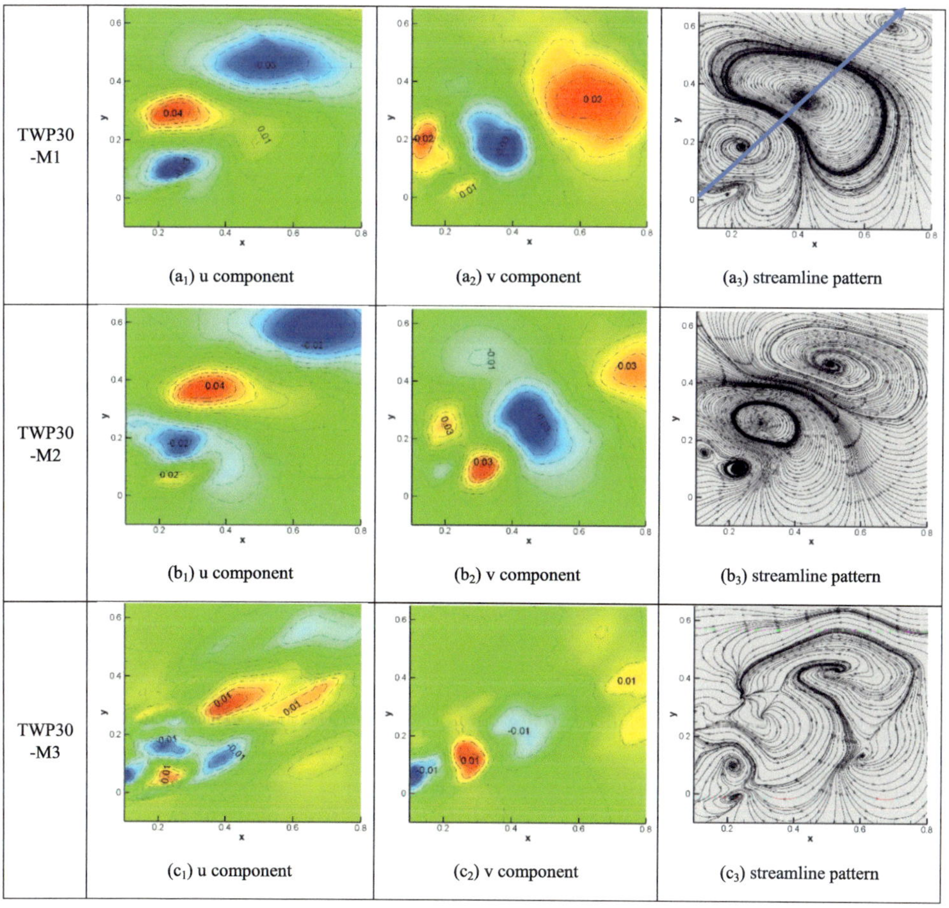

Figure 16.
The first three POD mode patterns of the wake velocity field at horizontal plane of 1/6 H under TWP30 (a) contour of u component; (b) contour of v component; (c) streamline pattern.

submerged in the turbulent boundary layer. The POD method is employed to extract the spatial/spectral features of the pressure pattern and flow pattern, and thus provides a physical mode interpretation of the aerodynamic characteristics of tall buildings subject to twisted winds. The primary conclusions are as follows:

1. POD can successfully extract the pressure mode and flow mode as well as capture the energy and spectral features of the dominant modes under the influence of CWP and TWP. The difference between the coherent structure in the twisted flow and the conventional flow is revealed, and the root cause of the influence of the twisted wind on the wind force and aerodynamic correlation of the high-rise building model is explained from the perspective of mode pattern.

2. For the pressure pattern, the first mode has a similar distribution to the mean pressure field, and is found to correspond to the low-frequency shift mode. While the second mode reveals various vortex shedding phenomena (i.e., Karman vortex, LEVS, TEVS, AVES) for the 0°and 90° cases. Compared with CWP, TWP largely modifies the pressure distribution, the power spectral feature

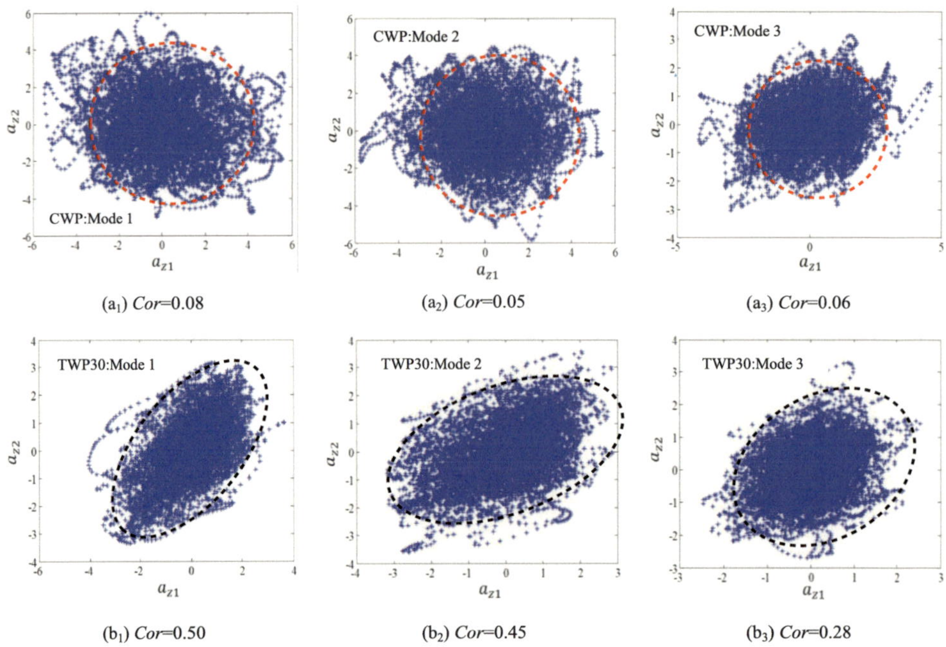

Figure 17.
Correlation of the first three POD mode coefficients of two horizontal planes at height of z = 1/6H and 1/2H for (a) CWP; (b) TWP30.

and the energy proportion of the dominant modes, and this modification in the POD mode pattern causes the twisted flow to have a significantly different impact on the aerodynamic properties of buildings. Specifically,

i. In the case of 0°AOA, when exposed to CWP, the first mode pattern has a similar distribution to the mean pressure field, implying that the first POD mode corresponds to the low-frequency shift mode. Whereas the second mode shape presents the asymmetrical feature, which possibly reflects the fluctuating properties induced by the periodic wake shedding. In the presence of twisted wind, the POD mode pattern on building surfaces is significantly modified, whose distribution appears more asymmetric and non-uniform. It is evident that the vertical varying wind direction has a more noticeable influence on the mode shape on the leeward surface than on the windward surface. Specifically, the first mode shape severely deflects towards the right side near the downstream region, while the second mode shape indicates that twisted flow could amplify the positive region near the left-side surface but shrink the negative area near the right-side surface. Such phenomena could be also observed for the second mode shape distributed on the side surfaces. This observation demonstrates that the conical vortex near the left side is greatly enhanced while it is largely suppressed near the right side, as highlighted by the black circle.

ii. In the case of 90°, under CWP, a shift mode is also observed for the first POD mode. Notably, the pressure pattern on the left and right sides are symmetrically distributed, but experience an initially upward and then downward process from the windward edge to the leeward edge. Such finding indicates that flow separation associated with flow reattachment simultaneously occurs at the side surface of a building with a large side ratio of 3:1. Similar to the 0° wind incident angle case, the asymmetrical second mode of CWP is highly related to the vortex shedding phenomenon. Affected by TWP, the continuous varying wind directions along the building height cause the reattachment location on the right-side surface moves forward, while that on the left-side surface basically disappears. Overall, the TWP has a similar influence on the mode shape of the windward and leeward surfaces for buildings attacked by 0°and 90° approaching wind.

3. For the flow pattern, its spatial–temporal features under CWP and TWP are successfully captured by POD. The main mode pattern of CWP is found to be controlled by the ILEVS and TEVS phenomenon, which corresponds to frequencies of f_{11} = 0.04 and f_{12} = 0.102 respectively. Whereas, the predominant POD mode of TWP is highly related to AEVS with a frequency of f_3 = 0.048. This modal pattern is reported to twist towards the wind direction with a more regular shape, even though it becomes less intense. Moreover, the spanwise correlation of the main POD mode at two planes of different heights is also significantly enhanced by TWP.

The study has comprehensively revealed the potential mechanisms of the twisted-wind effect that underlies the pressure field and flow field of tall buildings exposed to TWP. The mode interpretation of aerodynamic characteristics is expected to shed light on the fluid-induced interaction between twisted flows and tall buildings, and provides a better understanding of the underlying mechanism of the twisted-wind effect. The obtained conclusions facilitate useful references for designing and constructing tall buildings in hilly terrain where twisted wind occurs frequently, and more importantly, for TWP-induced vibration control measures.

Acknowledgements

The work described in this paper was supported by a grant from the Research Grants Council of the Hong Kong Special Administrative Region, China (Project No. 16207118 and 16207719), which is gratefully acknowledged.

Nomenclature

AOA	Angle of attack
ABL	Atmospheric boundary layer
CWP	Conventional wind profile
TWP	Twisted wind profile
MTA	Maximum twisted angle at ground level
POD	Proper orthogonal decomposition
ROMs	Reduced-order models
SMPSS	Synchronous multi-pressure sensing system

Author details

Lei Zhou and Kam Tim Tse*
Department of Civil and Environmental Engineering, The Hong Kong University of Science and Technology, Clear Water Bay, Kowloon, Hong Kong, China

*Address all correspondence to: timkttse@ust.hk

References

[1] Weerasuriya AU, Hu ZZ, Zhang XL, Tse KT, Li S, Chan PW. New inflow boundary conditions for modeling twisted wind profiles in CFD simulation for evaluating the pedestrian-level wind field near an isolated building. Building and Environment. 2018a;**132**:303-318

[2] Weerasuriya AU, Tse KT, Zhang X, Kwok KCS. Equivalent wind incidence angle method: A new technique to integrate the effects of twisted wind flows to AVA. Building and Environment. 2018b;**139**:46-57

[3] Tse K-T, Weerasuriya AU, Zhang X, Li S, Kwok KCS. Pedestrian-level wind environment around isolated buildings under the influence of twisted wind flows. Journal of Wind Engineering and Industrial Aerodynamics. 2017a;**162**: 12-23

[4] Tse K-T, Weerasuriya AU, Zhang X, Li SW, Kwok KCS. Effects of twisted wind flows on wind conditions in passages between buildings. Journal of Wind Engineering and Industrial Aerodynamics. 2017b;**167**:87-100

[5] Liu Z, Zheng C, Wu Y, Flay RGJ, Zhang K. Investigation on the effects of twisted wind flow on the wind loads on a square section megatall building. Journal of Wind Engineering and Industrial Aerodynamics. 2019a;**191**:127-142

[6] Liu Z, Zheng C, Wu Y, Flay RGJ, Zhang K. Wind tunnel simulation of wind flows with the characteristics of thousand-meter high ABL. Building and Environment. 2019b;**152**:74-86

[7] Feng C, Gu M, Zheng D. Numerical simulation of wind effects on super high-rise buildings considering wind veering with height based on CFD. Journal of Fluids and Structures. 2019;**91**:102715

[8] Zhou L, Hu G, Tse KT, He X. Twisted-wind effect on the flow field of tall building. Journal of Wind Engineering and Industrial Aerodynamics. 2021a;**218**:104778

[9] Zhou L, Tse KT, Hu G, Li Y. Higher order dynamic mode decomposition of wind pressures on square buildings. Journal of Wind Engineering and Industrial Aerodynamics. 2021;**211**:104545

[10] Zhou L, Tse KT, Hu G, Li Y. Mode interpretation of interference effects between tall buildings in tandem and side-by-side arrangement with POD and ICA. Engineering Structures. 2021b;**243**: 112616

[11] Ding F, Kareem A. Tall buildings with dynamic facade under winds. Engineering. 2020;**6**:1443-1453

[12] Nankai K et al. Linear reduced-order model based on PIV data of flow field around airfoil. Transactions of the Japan Society forAeronautcal and Space Sciences. 2019;**62**:227-235

[13] Kareem A. Mapping and synthesis of random pressure fields. Journal of Engineering Mechanics-Asce. 1989;**115**: 2325-2332

[14] Chen XZ, Kareem A. Proper orthogonal decomposition-based modeling, analysis, and simulation of dynamic wind load effects on structures. Journal of Engineering Mechanics. 2005; **131**:325-339

[15] Sirovich L. Turbulence and the dynamics of coherent structures. 1. Coherent Structures. Quarterly of Applied Mathematics. 1987;**45**:561-571

[16] Baker CJ. Aspects of the use of proper orthogonal decomposition of

surface pressure fields. Wind and Structures. 2000;**3**:97-115

[17] Bastine D et al. Stochastic wake modelling based on POD analysis. Energies. 2018;**11**(3):612

[18] Carassale L, Brunenghi MM. Statistical analysis of wind-induced pressure fields: A methodological perspective. Journal of Wind Engineering Industrial Aerodynamics. 2011;**99**:700-710

[19] Tamura Y et al. Proper orthogonal decomposition of random wind pressure field. Journal of Fluid Structure. 1999;**13**: 1069-1095

[20] Wang F, Lam KM. Geometry effects on mean wake topology and large-scale coherent structures of wall-mounted prisms. Physics of Fluids. 2019;**31**:125109

[21] Wang HF, Cao HL, Zhou Y. POD analysis of a finite-length cylinder near wake. Experiments in Fluids. 2014;**55**: 1790

[22] Tse K-T, Weerasuriya AU, Kwok KCS. Simulation of twisted wind flows in a boundary layer wind tunnel for pedestrian-level wind tunnel tests. Journal of Wind Engineering and Industrial Aerodynamics. 2016;**159**:99-109

[23] Yu Y, Yang Y, Xie Z. A new inflow turbulence generator for large eddy simulation evaluation of wind effects on a standard high-rise building. Building and Environment. 2018;**138**:300-313

[24] Van Doormaal JP, Raithby GD. Enhancements of the SIMPLE method for predicting incompressible fluid flows. Numerical Heat Transfer. 1984;**7**: 147-163

[25] Deniz S, Staubli T. Oscillating rectangular and octagonal profiles: Interaction of leading-and trailing-edge vortex formation. Journal of Fluids and Structures. 1997;**11**:3-31

[26] Zhang H, Xin D, Ou J. Wake control using spanwise-varying vortex generators on bridge decks: A computational study. Journal of Wind Engineering and Industrial Aerodynamics. 2019;**184**:185-197

[27] Schmid PJ. Dynamic mode decomposition of numerical and experimental data. Journal of Fluid Mechanics. 2010;**656**:5-28

[28] Zhang Q, Liu Y, Wang S. The identification of coherent structures using proper orthogonal decomposition and dynamic mode decomposition. Journal of Fluids and Structures. 2014; **49**:53-72